MARITIME UNMANNED

MARITIME UNMANNED

FROM GLOBAL HAWK TO TRITON

ERNEST SNOWDEN
AND **ROBERT F. WOOD JR.**

Naval Institute Press
291 Wood Road
Annapolis, MD 21402

Library of Congress Cataloging-in-Publication Data

Names: Snowden, Ernest, date, author. | Wood, Robert F., author.
Title: Maritime unmanned : from Global Hawk to Triton / Ernest Snowden, and Robert F. Wood, Jr.
Other titles: From Global Hawk to Triton
Description: Annapolis, MD : Naval Institute Press, [2021] | Includes bibliographical references and index.
Identifiers: LCCN 2021019767 (print) | LCCN 2021019768 (ebook) | ISBN 9781682477007 (hardcover) | ISBN 9781682477182 (pdf) | ISBN 9781682477182 (epub)
Subjects: LCSH: Triton (Drone aircraft)—History. | Global Hawk (Drone aircraft)—History. | United States. Navy—Procurement. | United States. Navy—Aviation—History.
Classification: LCC UG1243 .S64 2021 (print) | LCC UG1243 (ebook) | DDC 623.74/697—dc23
LC record available at https://lccn.loc.gov/2021019767
LC ebook record available at https://lccn.loc.gov/2021019768

♾ Print editions meet the requirements of ANSI/NISO z39.48-1992 (Permanence of Paper). Printed in the United States of America.

29 28 27 26 25 24 23 22 21   9 8 7 6 5 4 3 2 1
First printing

I am deeply indebted to my wife, Joan Snowden, and daughter, Hayley, for their suggestions, patience, and positivity, which propelled me forward in the drafting of this narrative and gave me the encouragement to set this important history to words.

◆ ◆ ◆ ◆ ◆

To my wonderful wife, Noreen Powell, who supported me throughout my entire aerospace and defense career and endured my long absences and many long hours at work, especially while I was focused on winning the Broad Area Maritime Surveillance contract for Northrop Grumman. She has always been there for me in every facet of my professional career. I could not have written this book without her as my trusted wingman.

◆ ◆ ◆ ◆ ◆

To the men and women of our United States Navy and the maritime patrol community, we salute your dedication and service to defending our great nation from the sea. We will always be in awe of your courage, commitment, and selfless sacrifice, always standing the watch so we may be free.

CONTENTS

ACRONYMS AND TERMS

ACTD   Advanced concept technology demonstration: a program initiated in 1994 to accelerate introduction of mature technologies by early involvement of warfighters following a focused demonstration program to enable investment decisions before commitment to a formal acquisition program; replaced by joint concept technology demonstration in 2006.

AESA   Active electronically scanned array: a type of phased array radar antenna that relies on electronic directional steering of the radio beam in lieu of mechanical steering.

A/F-X   Attack fighter aircraft experimental: originally established as the attack aircraft experimental (AX) program, a multirole attack/fighter for the Navy and potentially the Air Force in the wake of the cancelled advanced tactical attack aircraft program in 1991. When the Navy variant of the Air Force advanced tactical fighter was cancelled, Navy requirements were transferred to AX, resulting in the name change to A/F-X. The Office of the Secretary of Defense cancelled the A/F-X program in the Bottom-Up Review in 1993 due to anticipated excessive cost and risk.

ANA   Association of Naval Aviation: a professional association dedicated to furthering understanding of the role and contribution of naval aviation as a component of the national defense.

APL   Applied Physics Laboratory: a research center affiliated with the Johns Hopkins University that operates as an independent, non-profit organization conducting research, development, and systems engineering to support national security needs.

ASW   Antisubmarine warfare: a branch of underwater warfare concerned with detecting, tracking, and attacking enemy submarines.

ATA   Advanced tactical aircraft: a long-range, very-low-observable, high-payload medium-attack aircraft proposed to replace the

Grumman A-6 in the carrier-based medium-attack role, begun as a competitive acquisition program in 1983. When the team of McDonnell-Douglas and General Dynamics was selected to develop the ATA, it was designated A-12 and named Avenger II. The secretary of defense cancelled the program in 1991 for exceeding weight, cost, and schedule targets.

ATDC Advanced technology development center: a unit that existed in the 1990s in Northrop Grumman's integrated systems sector that was charged with investment and capture of front-end new business in advanced technology.

AX Attack aircraft experimental: a joint aircraft development program initiated in the early 1990s with participation by the Navy and the Air Force to replace strike aircraft nearing the end of their service lives. The AX was to replace the Navy A-6 and the Air Force F-111, F-15E, and F-117, but it was cancelled in late 1993.

BAA Broad area announcement: a process technique used by U.S. government agencies to solicit proposals from outside groups for research and development projects that are usually not expected to produce end-item hardware but rather to advance the state of the art in technology.

BAMS Broad area maritime surveillance: the program name of an initiative to provide the Navy's maritime patrol community with a persistent global intelligence, surveillance, and reconnaissance system capable of detecting, tracking, classifying, and identifying maritime and littoral targets.

BAP Business acquisition process: a set of instructions and protocols established by businesses to guide efforts toward obtaining new business.

BRAC Base realignment and closure: a process created by congressional statute in 1988 to empower a commission to recommend for closure U.S. government facilities deemed inefficient or excess to the military, thereby mitigating the undue influence of pork-barrel politics when those facilities were named.

CE Concept exploration: formerly the initial phase of Department of Defense system acquisition, wherein an acquisition strategy was developed and alternatives assessed in response to a mission need statement.

CNA   Center for Naval Analyses: a federally funded research and development center that performs directed research and analysis for the U.S. Navy, Marine Corps, and other Department of Defense organizations.

CNO   Chief of Naval Operations: the uniformed head of the U.S. Navy who serves as the Navy member of the Joint Chiefs of Staff in the Department of Defense and advises the president on naval matters.

CSA   Common support aircraft: initiated in 1993, the CSA concept was to evaluate the possibility of developing a single carrier-borne aircraft that could replace the E-2C for the airborne early warning mission, the C-2A for the carrier on-board delivery mission, the S-3B for carrier-based antisubmarine warfare, and the ES-3A for carrier-based electronic and signals intelligence collection. A common airframe was envisioned as the foundation for all four missions, with specific equipment and different avionics adapted to the airframe for each mission. By 1999 the difficulty of using a single airframe to host widely disparate missions was becoming apparent as estimated costs were rising rapidly. The concept was allowed to lapse, and further activity was discontinued by 2000.

DARO   Defense Airborne Reconnaissance Office: established in 1993, DARO was organized and chartered to provide increased senior management focus on the development and procurement of defense-wide airborne reconnaissance capabilities, including manned and unmanned aircraft, sensors, and data links. The defense authorization act of fiscal year 1998 transferred most oversight authority of DARO back to the respective military services.

DARPA Defense Advanced Research Projects Agency: an agency of the Department of Defense responsible for the development of emerging and transformational technologies for use by the military by working with science and technology centers in academia, industry, and government.

DSB   Defense Science Board: an advisory committee to the Department of Defense founded in 1956 reporting to senior leaders on complex technology issues confronting the department in research, engineering, and manufacturing, in combination with strategy, tactics, and operational concepts to address those issues. The DSB comprises forty-eight members and several ex officio members from the science and technology community at large, including the chairs of the military services and of various defense advisory committees. The DSB today reports through the under secretary of defense for acquisition technology and logistics.

EN   Evaluation notice: a notification from a government customer in the form of a question regarding some aspect of a proposal that has been submitted. The EN normally requires a formal response from the company submitting the proposal.

ETOS   Effective time on station: a measure of the ability of a system to remain operational in providing surveillance coverage.

FBE   Fleet battle experiment: an evaluation of new concepts and technologies that normally takes place inside a scheduled fleet exercise in order to inject these technologies and concepts into a more realistic operational setting under real-world conditions and thereby assess which technologies and concepts proved workable and what follow-on experimentation should be pursued.

GHMD   Global Hawk Maritime Demonstrator/Demonstration: a program established and managed by the Navy in conjunction with the Air Force Block 10 Global Hawk program phase that added maritime modes to the existing synthetic aperture radar system as well as dual-band data link and enhanced electronic support measures in order to evaluate a high-altitude unmanned aerial vehicle adjunct to the Navy's multimission maritime aircraft replacement to the P-3C. GHMD consisted of two air vehicles, two launch and recovery elements, one mission control element plus Kurz under satellite and data terminals, as well as spares and support equipment. GHMD aircraft began operating in 2007 from Naval Air Station Patuxent River, Maryland, entering multiple fleet exercises and experiments.

HAE UAV   High-altitude, long-endurance unmanned aerial vehicle: a capability to be acquired by means of an advanced concept technology demonstrator (ACTD), begun in 1994 by the Defense Advanced Research Projects Agency and the Defense Airborne Reconnaissance Office to develop and demonstrate high-flying UAV systems capable of affordable, continuous, all-weather, wide-area surveillance. A stated goal of the HAE UAV ACTD was to avoid previous frustrations with UAV development programs by exploring ways to simplify and accelerate the acquisition process. The HAE UAV ACTD program yielded two separate classes of vehicle: the Tier II+ Global Hawk and the Tier III– Dark Star.

ISR   Intelligence, surveillance, and reconnaissance: a term broadly defining the collection, processing, and dissemination of information relevant to military commanders, normally obtained by the dispositioning of sensors, networks, communications, and human resources to gather and interpret knowledge of activities, areas, or targets of interest.

JEFX   Joint expeditionary force experiment: an Air Force exercise established in the aftermath of Operation Desert Storm to evaluate new technologies and concepts in a simulated battle environment that emulated real-world operational settings. As other services were invited to participate, the scope expanded to regular joint participation. Fiscal constraints resulted in the discontinuance of the JEFX event in 2013.

MFAS   Multifunction active sensor: an electronically scanned array radar designed for maritime surveillance.

MMA   Multimission maritime aircraft: a major defense acquisition program to provide a replacement aircraft system for the aging P-3C aircraft to perform antisubmarine and anti–surface warfare, maritime targeting and strike, and maritime intelligence, surveillance, and reconnaissance for the maritime patrol and reconnaissance force. MMA is intended to meet the broad area maritime and littoral armed intelligence, surveillance, and reconnaissance mission need statement validated by the joint requirements oversight council in February 2000. In June 2004 Boeing was selected for design and development of the P-8 aircraft—a militarized version of the Boeing 737-800—for the MMA requirement.

MTS-B   Multispectral targeting system: an airborne sensor system that combines electro-optic/infrared, laser designation, and laser illumination of potential targets; "B" refers to its place in a series of MTSs produced sequentially.

NAVAIR/NAVAIRSYSCOM

Naval Air Systems Command: reporting to the Secretary of the Navy through the assistant secretary for research, development, and acquisition, NAVAIR is the single command charged with developing, procuring, and supporting aircraft, weapons, and aviation systems. NAVAIR employs more than 27,000 technically trained civilians and 1,500 active-duty military personnel in locations across the United States.

NDIA   National Defense Industrial Association: an industry advocacy organization comprised of geographically dispersed chapters and divisions with 1,600 corporate and 85,000 individual members. It is a nonpartisan, nonprofit, educational association founded to educate its constituencies on all aspects of national security.

NWDC   Navy Warfare Development Command: originally established in 1998 as part of the Naval War College in Newport, Rhode Island, before relocating to Norfolk, Virginia, as part of the Fleet Forces Command, the organization is lead agent for Navy warfighting concept generation and development. It is charged with identifying innovative solutions to emerging threats, developed through experimentation, modeling and simulation, and lessons.

OTA   Other transaction authority: a term denoting the Department of Defense's authority, granted under 10 U.S. Code 2371, to award prototype projects that show relevance to weapons systems to be acquired by the department. Not covered by federal acquisition regulations, the added flexibility OTs offer as business tools can support more efficient, lower risk acquisition. Each service may exercise the authority up to $250 million without approval from the under secretary of defense for acquisition and logistics, provided at least one nontraditional defense contractor is significantly involved in the project.

PBD   Program budget decision: in the planning, programming, and budget execution process, the program budget decision documented issues arising during the review of the service program objective memoranda by the Office of the Secretary of Defense and reflected decisions as to the appropriate program and funding levels that should go forward in the annual budget submission. Normally, the military service would be provided a draft PBD for comment or reclama before being signed by the secretary of defense. The PBD was superseded by the resource management decision document in 2012.

POM   Program objective memorandum: a comprehensive plan prepared by the military services and defense agencies for how they will allocate funding for programs to satisfy the defense planning guidance issued by the Office of the Secretary of Defense. The POM is part of the programming phase of the planning, programming, budgeting, and execution process. The POM covers a five-year window with the services' intent for how they will balance their programs given the available budget. It is normally submitted to the Office of the Secretary of Defense in late summer/early fall and is accompanied by an update to the future years defense plan.

PUMAS   Persistent unmanned maritime airborne surveillance: a Navy competitive program undertaken in 2006 to obtain from the U.S. aerospace industry an independent assessment of the overall utility of the Navy's evolving manned and unmanned maritime intelligence, surveillance, and reconnaissance (ISR) systems, circa 2013. The study assessments were based on a specific set of Navy-provided scenarios and industry team–identified airborne ISR gaps.

QDR   Quadrennial Defense Review: a review of Department of Defense strategy and priorities mandated by congressional legislation every four years. The QDR assesses threats and challenges that the nation faces and rebalances the department's strategies, capabilities, and forces to address current conflicts and counter future threats.

QFD   Quality functional deployment: a technique that translates customer requirements—expressed in the customers' language—into an action plan. QFD will establish critical customer attributes and measure or rank their importance. For Navy applications, it is used as a means of compiling and synthesizing communication from fleet users regarding their tactical or mission needs and transcribing those into a set of discrete capabilities.

RFP   Request for proposal: the document that solicits a proposal from a prospective contractor in response to a set of technical and cost parameters.

RMA   Revolution in military affairs: the transformation of warfare by information-age technologies such as computers, microelectronics, and precision weapons. Information superiority is the enabling cornerstone of RMA, defined as the capability to collect, process, and disseminate an uninterrupted flow of information while exploiting or denying an adversary's ability to do the same.

SWAP/SWAP-C
       Size, weight, and power/size, weight, and power-cooling: generally used in reference to the design compromises required to arrive at optimal sizing of an aircraft or airborne system's dimensions, weight, power, and cooling required.

TAGS   Tactical auxiliary ground station: a Navy-owned secure facility on site at Naval Air Station Patuxent River, Maryland, with the capabilities to downlink and process imagery and other intelligence, surveillance, and reconnaissance data from the sensors on the BAMS demonstrator aircraft and disseminate it to Navy and other external organizations in the United States and to U.S. subscribers in overseas locations.

VP   Navy notation for a maritime patrol squadron, typically engaged in surface surveillance, antisubmarine warfare, and anti–surface warfare. Patrol squadrons are denoted by one or more numerals following the letters VP (e.g., VP-11, or Patrol Squadron Eleven).

VQ   Navy notation for a fleet air reconnaissance squadron, typically engaged in airborne collection, processing, and analysis of electronic signals intelligence against threat or hostile forces. Now deactivated, VQ squadrons were denoted by a numeral following the letters VQ (e.g., VQ-1, or Fleet Air Reconnaissance Squadron One).

INTRODUCTION

This is a story of monumental success against long odds. It equally is an exploration of how bold ideas, once adopted, can disrupt conventional practices and norms. It is a story that should remind the reader of the long struggle involved in taking aircraft to sea on aircraft carriers—a notion dismissed, even ridiculed, by most Navy leaders at the beginning of the last century. Then, only through the perseverance of a few fierce, forward-thinking zealots was the potential realized for increased reach, precision, and versatility by the operation of air wings from flight decks. When asked to cite the most important development in the history of warfare at sea, Adm. Marc Mitscher, commander of the Fast Carrier Task Force in the Pacific in World War II, was quick to reply that *pilots* had become "the most important thing in battle."[1]

If that question could be asked today, Mitscher would likely be no less certain of the central importance of the pilot. More than seventy-five years later, the role of the naval aviator in the Navy's identity remains paramount. In that time, though, technology has reshaped perceptions regarding many accepted patterns or behaviors. For example, when Mitscher spoke, more than a quarter-million telephone switchboard operators were employed across the United States manually connecting incoming and outgoing call lines. Today, automatic telephone switching machines have replaced those jobs wholesale. That digital technology can assume more of the physical human activity required for most jobs has become an inescapable fact, even though cultural resistance still militates against its widespread and immediate embrace in many sectors of our society.

Naval aviation remains a pilot-centric enterprise, but how we think about aircrew and their physical presence as inseparable from their aircraft is beginning to give way. The central idea of our story is that the first important step in naval aviation of breaking with the traditional view, of departing from the accepted cultural norm, was exemplified by the long struggle and ultimately successful effort to bring the MQ-4C Triton unmanned aircraft into being. The narrative begins its arc in the early 1990s, when the U.S. military services were reassessing their missions and equipment in a post-Soviet world. The focus narrows to the Navy's replacement of aging maritime patrol aircraft starting at the end of the 1990s and concluding at the end of the first decade of the 2000s.

This story begins, oddly, in the U.S. Air Force. In that service's search for a less expensive way to perform airborne surveillance in the 1990s, some saw unmanned aircraft such as the Global Hawk as the logical way to remove cost—and not coincidentally, risk to the pilot—from the mission. For the Air Force as a whole, an organization steeped in the culture of the pilot and his machine, institutional resistance nearly sidelined the Global Hawk on multiple occasions. Migrating the Air Force Global Hawk into front-line Navy service, crossing one culture and insinuating into another, took extraordinary vision and leadership. We see that brought to bear by a few iconoclasts in uniform and in industry who worked to override the Navy's own resistance to displacing the airborne pilot with an unmanned machine. In its first study of the Air Force Global Hawk in 2000, the Navy, to create some separation for its own effort, applied a more familiar moniker to the unmanned aircraft and its associated ground control equipment: broad area maritime surveillance (BAMS). This would remain the popular title until the program achieved a threshold of maturity, warranting a more formal title: Triton.

To better appreciate the intricate maneuvering involved to make all that happen, the reader is immersed in the detail of the sometimes sluggish, always restrictive, defense acquisition processes. To understand how a defense company organized its resources and strategized its application in response to the acquisition process also requires grounding in industry processes. To aid the reader in understanding outcomes, the authors have drawn upon their intimate familiarity with the story to inject lessons learned and to set forth repeatable steps applicable to a wide range of competitive situations. These looks deep inside acquisition rules are foundational to seeing the story as a whole: how the inertia of culture and bureaucracy were overcome by smart strategic choices to break through. The story and long path behind the industry strategy in creating the BAMS capability is an example of the best attributes of aeronautical technical competence, marketing ingenuity, perseverance, and customer support. The result: an alliance of innovators and disrupters, willing to challenge staid, entrenched attitudes with a better idea, has revolutionized the way maritime surveillance is performed.

ORIGINS

The U.S. Navy has, throughout its history, been circumscribed by culture. Its internal warfighting communities have built esprit de corps and cohesion by adopting values, behaviors, and practices that then become ingrained. The pattern of their training, equipage, operations, and tactics becomes deeply rooted in a shared history that makes for a reliable predictor of how they approach and engage with the outside. Too often, when those cultural norms endure unchanged for extended periods, they can become sclerotic, unresponsive to the arrival of new and relevant advancements; worse, the culture could militate against fresh thinking, the introduction of needed updates to operations, and, especially, the adoption of new technologies. Examples abound: the slow transition from sail to steam in the mid-1800s, the service's disdain for the submersible at the beginning of the twentieth century, or the barely lukewarm reception for the earliest airplanes a generation later.

Nearing the close of its first decade, U.S. naval aviation in 1920 was still considered by most in the service to be an inventive oddity or a dubious appendage to the heavy-gunned battle line. The Navy of the early 1920s was "tradition-bound and overly conservative. Its reluctance to break from a doctrine centered on decisive engagements between battleship-centered fleets"[2] suggested a naval culture wedded to practices and technologies already mature at the turn of the century and beginning to ossify as newer technologies were rapidly appearing. No less a personage than the Chief of Naval Operations, Adm. William Benson, was heard in 1919 to say, "The Navy doesn't need airplanes. Aviation is just a lot of noise."[3] By the late 1920s, though, aviation was beginning to insinuate itself into a prominent place in the planning and conduct of the Fleet Problems. These large-scale maneuvers consumed most of the surface Navy and served to operationalize new technologies and employment concepts throughout the 1920s and 1930s. As aircraft performance improved during those years and aircraft carriers became more central to the exercise outcome, adherents of aviation changed the discourse about its role and altered long-held perceptions at all levels of the service. The Fleet Problems ran their course and by the 1940s were superseded by the real problem of meeting and defeating enemy naval forces at sea. In the years that followed, other wars had the effect of hastening capability advances to meet urgent combat needs. However, as the pace of technological advance quickened over the last fifty years, it became increasingly clear that what was needed was an ordered, repeatable mechanism for rationalizing and prioritizing capability needs and technology investment.

In our time, an orderly process has evolved to replace this more ad hoc behavior for recognizing and adopting capability advancements. Like so much process that originates in the minds of Pentagon bureaucrats, it comes to us in convenient shorthand that supplants spoken English and becomes the lingua franca of the defense establishment. The shorthand we know as acronyms might best be understood as the English subdialect spoken offhandedly and naturally in the halls of the Pentagon and in military acquisition directorates. Much of the narrative in our story relies on acronyms; however, the authors have made every effort to minimize their use by keeping the "Pentagonese" in longhand. A prime example might be the acronym JCIDS—the joint capabilities integration and development system—coined to describe the orderly process by which the military services articulate their needs. Imposed by statute in 2003 and shaped by regulation, JCIDS standardizes the means for the Navy and the other services to identify and prioritize the capabilities each requires to meet its warfighting obligations. JCIDS is the requirements generation process and is intended to set in motion the "pull" for enabling technologies as well as instigating changes in doctrine and operating concepts. The Navy maritime patrol and reconnaissance community's requirement for a replacement aircraft in the late 1990s, for example, resulted in the multimission maritime aircraft requirement following a process not unlike JCIDS, even though it was formulated just a few years before JCIDS was officially issued.

Less conventional and often serendipitous are the means by which disruptive new technologies are pushed to the fore and adopted into regular service use to fulfill a warfighting requirement that might not yet have been defined. High-profile examples do exist: the Internet, global positioning systems, night vision, and lasers. But where a process for requirements pull has become assimilated into the bureaucratic norm, technology push more typically must confront entrenched cultural bias and preference for accepted and familiar equipment, operating concepts, tactics, training, and, in some cases, even the self-image or warrior ethos of those who populate a service community. For disruptive technologies to be recognized and adopted, culture often becomes the impediment around which visionaries must organize resources and strategy to overcome.

Culture—or more precisely, the collision of cultures new and already established—will inspire leaders who step into the resulting tumult to create order, organize, and drive teams toward an intended purpose. What follows is an exposition of leadership examples and strategy formulations that proved pivotal in first recognizing and then overcoming an entrenched cultural resistance to the introduction of an unmanned aircraft into a warfare mission that had been performed successfully since the beginning of naval aviation by manned aircraft.

In our narrative, the cultural impediment initially proved to be an unalterable design by an entire maritime patrol and reconnaissance community—operators, acquirers, sponsors, and advocates—to resolve its manned aircraft recapitalization dilemma by pursuing a one-for-one like-replacement, virtually the same capability but with turbofan jet engines. With that cultural single-mindedness subdued and overcome largely by the successful formulation and execution of business strategy, cultural resistance then arose within the corporation itself—due to the inability to merge disparate business units and then to form a cohesive and fiercely competitive team that could see past recent Global Hawk successes with the Defense Advanced Research Projects Agency and the Air Force to adapt to new conditions in a Navy contest against powerful corporate competitors.

In large bureaucracies—the U.S. Navy and the defense companies described herein—the emergence of the leader may ultimately have a transformational impact on an enterprise. In the foregoing example of naval aviation in its formative years, the prevailing disdain for the new technology could not have been reshaped without the energies and vision of two classes of uniformed aviation advocates that emerged in that first decade to take on the culture, even at the risk of shortening their careers. At the more senior level were the inspirational leaders—officers such as Rear Adm. William Moffett, chief of the Bureau of Aeronautics—who grasped the importance of aviation in the Navy and committed themselves to organizing the people and resources, lobbying Congress, and cajoling peers. Moffett's milieu was strategy; from long experience and instinctual feel for the service's needs, he architected a way to appropriate resources to attain a best outcome. At the lower rung were the disrupters, those officers who commanded squadrons and the early aircraft carriers—among them Capt. Ken Whiting—who were instrumental in working out the first techniques and protocols for aviation operations at sea. The uniformed disrupters were joined by an iconoclastic few civilian designers, engineers, and aeronauts—entrepreneurs such as Glenn Curtiss—who worked tirelessly to make their designs work for a Navy requirement and to press the Navy's bureaus and Congress for acceptance.

Seventy years on, a new technology was roiling the staid culture of the maritime patrol and reconnaissance community, a warfighting group still attached to the heavy multi-engine, propeller-driven, large-crewed patrol planes that now needed replacement. Unmanned aerial vehicle (UAV) prototypes were just beginning to show promise, and with a degree of maturity that suggested their potential usefulness in the maritime mission. Yet it would again take the commitment of a few inspirational leaders, and the exertions of disrupters both uniformed and civilian, to bend a culture to accept and eventually embrace the new technology.

As UAV technology was gaining new adherents in the 1990s, Secretary of Defense Les Aspin's "Last Supper" unleashed a wave of industry consolidations. Described in a media report that made reference to the biblical last supper, Aspin, early in his tenure, convened a dinner with more than a dozen chief executives of major defense companies to announce the end of the old order of the defense-industrial relationship. He sermonized about the start of a new order that would necessarily involve fewer companies competing for reduced shares of a much smaller defense budget. The Cold War between the United States and the Soviet Union had symbolically just ended with the fall of the Berlin Wall, and defense budgets were accordingly going into steep decline. Marquee companies with long histories in aircraft design and manufacture simply disappeared, most acquired by a few survivors yearning for economies of scale or eager to harvest already-awarded government contracts. Here, too, culture was a crucial factor in how the survivors were able to absorb the people, legacy resources, and customer relationships of the acquired businesses. Just as the Navy was (and is) circumscribed by culture, so too were these defense companies, and the behavioral norms inside these newly consolidated corporations were sometimes in stark, disagreeable juxtaposition. As will be explained, this would unduly impede early attempts to advance UAV technology. Once again, inspirational leaders and disrupters would rise above, able to frame issues, develop common purpose, harness the best of the constituent cultures, and create winning action plans.

Entering the first year of the twenty-first century, the stage was set for migrating a Global Hawk unmanned aircraft derivative from the Air Force to the Navy, a process that would ultimately take more than twenty years from the initial concept study to an operational capability recognized as the condition of an all-up squadron ready for deployment to conduct a search orbit. The leadership to effect this transition necessarily emerged initially at a corporate business enterprise level for kicking off the strategy and pursuit process—in our story, at the Northrop Grumman Corporation, in its profit-and-loss unit inside its integrated systems business sector. The stakes were high in overcoming service bias initially to gain acceptance of an unmanned adjunct to their maritime patrol and reconnaissance mission; for Northrop Grumman, the ability to gain a Navy customer in addition to the Air Force was key to growing the Global Hawk business franchise. The stakes were higher, in fact, in the subsequent Navy competition for a UAV-centered concept that came to be known initially as broad area maritime surveillance; the mandate for leadership insight and buy-in rose quickly to the sector and corporate levels for frequent review, advice, and approval. A strategic play involving substantial corporate investment for proposal and capital funds to preserve a growing business franchise and to keep major competitors from gaining a foothold in the high-altitude, long-endurance UAV market required engagement at the most senior level of the corporation. In an era in which the Department of Defense presented a dwindling number of large-ticket opportunities, a large competitive capture such as Navy BAMS could be a game-changer. This would be true not only for the corporate bottom line, but also for the Navy's emerging operational concept for maritime patrol and reconnaissance. Additionally, markets could be shrinking and/or overlapping with other businesses internal to a company or with competitors, so careful, detailed, and aggressive strategic thought was critical to even having a chance of winning.

All senior levels of Northrop Grumman leadership were involved throughout all phases of the long BAMS acquisition but especially near the end of the final proposal submittal and continuing through the protest by Lockheed Martin. There was no substitute for total leadership engagement on a critical competitive pursuit that BAMS represented, recognized at its inception as a game-changer for the company. Critical strategic decisions were sometimes made where all C-suite leaders were not in agreement on what to do. In the Navy BAMS competition, the leadership of the appointed capture executive—that individual given final responsibility and accountability for winning the competition—would prove a deciding factor.

Early planning, customer shaping, and defining the competitor domain were critical to setting the foundation for a winning strategy. The general elements of win strategy formulation governed all further strategy discussions up and down the leadership chain in the company. Those elements included but were not limited to customer value analysis (identifying "table stakes" and competitive discriminators), competitor analysis (developing a strengths, weaknesses, opportunities, and threats analysis for each and identifying their likely win strategy), early win strategy development (customer value analysis and trade-offs with a value proposition), strategy assessment (identifying critical assumptions of the early win strategy and those underlying the competitors' win strategies, and formulating critical questions for engaging the customer), and strategy refinement, early pricing, and updates to the competitor analysis.

In the pages that follow, we illuminate how the U.S. Navy's Triton UAV came to be. From its inception as BAMS to its operational realization as Triton, here is the story of bureaucratic challenges overcome, entrenched cultural biases turned, galvanizing leadership examples revealed, and capture strategies conceived and adapted to successfully anticipate a changing competitive environment. The story is told from the ground level—or, more appropriately to a Navy history, from a deckplate level—through the observations and experiences of participants, but with particular emphasis on the industry view. The initial BAMS concept would be augmented by a parallel prototyping of two proof-of-concept vehicles called Global Hawk Maritime Demonstrators (GHMD), which would evolve in time and through an early operational deployment to become known as the BAMS demonstrator (BAMS-D) by the Navy Fifth Fleet users. Finally, BAMS itself would earn its nom de guerre, Triton, in conformance with the Navy's tradition for naming all aircraft in the maritime patrol and reconnaissance family after characters from Greek mythology. From their personal involvement at different stages of the program's progression from BAMS to GHMD to BAMS-D and finally Triton, we contribute our own perspectives but also draw on the recollections, meeting minutes, correspondence, and personal notes of many colleagues with whom we worked closely throughout the nearly two decades required to get Triton into fleet service. When reassembling the history of a large defense acquisition, we must necessarily skirt the strictures on competitive-sensitive material and still cover the subject in enough detail to explain the competitive forces and capture strategies. We have made every effort to limit our narrative to sources in the public domain and to expose the manuscript to peer review for proprietary information.

This is a story of technology-driven breakthrough, inspired design innovation, industry-leading systems engineering competence, sensitivity to the operational needs of the maritime patrol community, and preeminent marketing strategy, aggregated to achieve a singular outcome in spite of sometimes determined opposition, widespread indifference, and the inertia of the unbending colossus that is the defense acquisition system. We do not intend to disparage any individuals, corporations, or military services. Any number of major defense acquisition programs over the last forty years experienced similar setbacks and achievements as the participants on those initiatives toiled to bring new technologies or concepts to the fore. Rather, by chronicling the tortuous acquisition history of Triton, we hope to underscore how several strains of principled, committed leadership were permitted to flourish inside an evolving corporate culture and, in parallel, a service culture, to attain a transcendent outcome: the mainstreaming of large-scale UAV technology in regular service inventory.

Those who contributed directly to Triton's ultimate success and who deserve mention here for their inputs, notes, and critiques of early drafts include Capt. Alan Easterling, USN (Ret.), Capt. Mark Turner, USN (Ret.), Dr. Maris Lapins, Mr. Howard Frauenberger, Mr. Dennis Hayden, Capt. Al Hutchins, USN (Ret.), and Col. James Tapp, USAF (Ret.), retired corporate vice president for business development at Northrop Grumman. Though not a direct contributor to this story, Mr. Rick Thomas' self-published book on the history of the Air Force UAV program titled *Global Hawk: The Story and Shadow of America's Controversial Drone* must be acknowledged as an authoritative source and inspiration for this story of the Global Hawk and its transition to U.S. Navy service.

The authors would be remiss in not acknowledging the careful and exacting scrutiny that Northrop Grumman members conducted on the final draft. Those members include Tim Paynter and Brian Humphreys. We are especially indebted to Tom Twomey at Northrop Grumman for shepherding the draft through a very rigorous review process. At the Naval Institute Press, the work of the editorial team deserves our sincerest thanks, as well as copyeditor Lisa Yambrick for her thoroughness in reviewing our text and making significant improvements. Finally, we are deeply indebted to Tom Cutler at the Naval Institute Press for his insights and for his perseverance in guiding the manuscript to final approval.

CHAPTER 1
INCEPTION

Last year we proudly celebrated the centennial of naval aviation. This year we've seen the roll out of a new patrol plane, the P-8. And now . . . the beginning of an unmanned tradition begins in our fleet with the unveiling of BAMS. History will record this introduction as a significant milestone in the second one hundred years of naval aviation.

—Adm. MARK FERGUSON, USN[1]

N avy tradition, dating to the first year of our declared independence as a nation, has affirmed special recognition for the commissioning pennant. Since those days, the pennant has taken the form of a long streamer flown from the masthead, heralding a ship's appointment as a registered vessel in Navy service. It conferred special obligations and responsibilities on the crew to maintain the ship in orderly condition in peace and in fighting trim for war. *Alfred* was the first American warship to be accorded the distinction of a commissioning pennant: "blue at the hoist, bearing seven white stars and single longitudinal stripes of red and white."[2] For one hundred years, until the sunset of the age of sail, pennants remained large by today's standards, often trailing more than fifty feet behind the mast. Modern ship appurtenances—radar dishes, radio antennas, and the like—necessitated a reduction in size to a more practical six-foot fly, even as its use was broadened ceremonially to include commissioned aviation squadrons.

So it was that on October 28, 2016, under high overcast and warm mid-fall temperatures in Jacksonville, Florida, a fresh pennant was broken for the commissioning of a newly formed aviation squadron, VUP-19. In a customary ritual, outwardly similar to many commissioning ceremonies that preceded it, guests and high-ranking visitors assembled in fold-down seats carefully arranged according to protocol inside hangar 117, facing a dais and lectern reserved for the presiding official and incoming commanding officer. The dominating presence of a full-scale aircraft replica on the tarmac, distinguished by flashes of squadron livery on the tail, with its nose protruding into the open hangar door, lent a visible touchpoint and reminder for the participants of the underlying purpose of the proceedings. This replica was a stand-in, a nonflying surrogate representing the imminent arrival of four fully functional aircraft that were authorized for squadron inventory. The squadron's flyable aircraft were still in the production pipeline at the manufacturer's facility, but the replica on static display bore enough detail to well serve the occasion.

What marked this ceremony as uniquely different, a singular departure from all maritime patrol plane squadron commissioning ceremonies that came before it, was that this squadron's aircraft, when delivered and made operational, would be unmanned. That October 2016 ceremony was historic in this sense: maritime patrol aircraft have been a fixture in the history of naval aviation since its beginnings prior to World War I and normally accommodated positions for a pilot, co-pilot, and flight engineer, navigators, tactical coordinator, sensor operators, intelligence specialists when the mission required, and, from time to time, ordnance experts. Each member of the crew specialized in some aspect of the mission depending on their particular function and training, but all played some part in the defining characteristic of patrol plane missions: long-duration over-water surveillance that required eyes on target—or on airborne sensor display—for detection or identification.

VUP-19 was, in fact, Unmanned Patrol Squadron 19, in whose service assigned naval aviators, naval flight officers, and enlisted aircrewmen would never leave the ground aboard their aircraft but would instead operate them remotely from a darkened control center. Buried deep inside a brick-and-mortar building festooned with large parabolic dish antennas, these operators would sit for long duty shifts bathed in blue light, monitoring sensor images and issuing digital commands from their keyboards hundreds if not thousands of miles removed from the aircraft they fly. The squadron's first aircraft, in fact, executed takeoff and landing from Naval Base Ventura County in Point Mugu, California, but were flown by pilots and mission operators sitting in their control center at Naval Air Station Jacksonville, Florida.[3]

On the tarmac about one hundred feet away from the ceremony that morning in Jacksonville was the Navy's primary manned patrol plane, the Boeing-produced P-8 turbofan jet-powered aircraft, by then set to enter its third year at full rate production and entering squadron service in increasing numbers. The P-8 Poseidon was so named to conform to the Navy's practice of assigning noms de guerre to its maritime patrol planes from Greek or Roman mythology, beginning with Neptune, then Orion, then Poseidon. In myth, Poseidon was one of the principal deities of the Greek pantheon whose domain encompassed the oceans. Poseidon's son, Triton—a half-man, half-fishtail sea god—would announce his father's comings and goings and would calm the seas by sounding a conch-shell trumpet. In much the same way that Triton heralded his father's movements and inclinations as a lesser god of the sea in Greek myth, the unmanned aircraft called to service with VUP-19 that October morning—as Triton—would serve as herald and caller for its larger manned aircraft progenitor, Poseidon.

But the relationship should not be interpreted as one of subordinate and senior. From inception, each aircraft has had distinct capabilities that permit the other to focus more on its strengths. Weapons carriage, for example was never envisioned for Triton, even in its earliest incarnation as a broad area maritime surveillance patrol craft. Some future ability to dispense sonobuoys was suggested in the original BAMS concept study, but nothing more lethal was contemplated: "Triton will provide a superior picture of what is happening above the surface, enabling Poseidon to focus on what is below the surface."[4]

From its inception nearly twenty years earlier as BAMS, the Northrop Grumman–produced Triton was proffered as the unmanned adjunct to the larger manned aircraft in a hybrid force, with each aircraft performing those maritime patrol and surveillance functions for which it was most adept but in a highly complementary, mutually supporting collaboration. Years later, the Navy would refine its description of Triton as a system for "persistent ISR [intelligence, surveillance, and reconnaissance] of nearly all the world's high-density sea-lanes, littorals and areas of national interest . . . designed to provide from five bases in both the continental United

States and outside the geographic perimeter of the U.S., near worldwide coverage through a network of orbits, or search patterns, with sufficient air vehicles to remain airborne for 24 hours a day, 7 days a week out to ranges of 2,000 nautical miles."[5]

Remarkably, that description from an unclassified fiscal year 2015 Navy budget highlight echoes almost word for word the summary report of Northrop Grumman's BAMS concept study produced fifteen years earlier, validating the soundness and exacting rigor of the original analysis. Even the operational deployment concept of five orbits was originally put forth in the Northrop Grumman study as five BAMS bases at Jacksonville, Florida; Kaneohe, Hawaii; Sigonella, Italy; Diego Garcia; and Kadena, Japan. At the time, those sites were also forward operating bases for the maritime patrol and reconnaissance community's deployed P-3 Orion turboprop aircraft. At the heart of that study was the Northrop Grumman assertion of a persistent surveillance coverage footprint obtainable when operating from those bases with sixteen to eighteen hours of time on station at a radius of three thousand nautical miles from base. This would become one of just a few pointed, defining discriminators that would turn Navy stakeholders into advocates in the early socialization of BAMS with Navy buyers and operators and, later, in the BAMS industry competition. The current basing concept as Triton fills out the Navy inventory is essentially unchanged, with the exception that Qatar replaced the proposed site at the now-inactive base at Diego Garcia and that operations were added at Point Mugu to expand the footprint with a second domestic U.S. site.

At the time of the BAMS concept study, the manned aircraft in this hybrid mix was yet to be officially decided—other aircraft manufacturers and commercial modification centers were staking out positions and shaping strategies for pressing a competitive advantage in a contest that was just getting under way for the multimission maritime aircraft (MMA). The MMA concept of replacing all land-based Navy support aircraft with a common new design airframe proved too ambitious and would soon narrow to a competition that would determine what path the Navy would follow to replace its rapidly aging fleet of about 250 P-3C Orion aircraft. Northrop Grumman theorized very early in the life of BAMS that with a sensor suite adapted to and specifically focused on the maritime environment and with connectivity established with a distant manned MMA aircraft, ship, or ground-based maritime control center, BAMS could provide the extended long-duration surveillance that would yield initial discovery, classification, and probable identification leading immediately to a contact alert—effectively a tripwire. An alert, in turn, would elicit an intercept response by the weapons-laden MMA "pouncer" aircraft for confirmation and prosecution of the contact.

That connectivity between the eventual manned aircraft MMA selectee, Poseidon, and eventual unmanned BAMS selectee, Triton, was finally demonstrated in June 2016 when the two aircraft successfully exchanged full-motion video for the first time inflight by means of a common data link (communications equipment that complies with a set of rules for imagery and signals intelligence describing standards for bandwidth and data transmission rates).[6] In the earliest description of the unmanned system, the MMA was hypothesized to actually manipulate and direct the unmanned BAMS aircraft's sensors to respond to unanticipated targets not already on the preprogrammed collection plan in what was termed level four control. That description leaned in the direction of ultimately extending the control tether to remotely establish level five control of the unmanned aircraft's flight plan itself, but from the manned MMA aircraft by operators on board in response to more urgent real-time mission retasking.

These were but a few of the operational adaptations that could be achieved, it was asserted, by augmenting the basic Global Hawk with a navalized set of sensors that could yield much improved results in surveillance of oceanborne targets. These assertions, put forth in Northrop Grumman's study, were contained in a Navy-contracted concept study in 2000. The first prototype Global Hawk had already flown two years prior as the so-called Tier II+ selectee in a Defense Advanced Research Projects Agency (DARPA) advanced concept technology demonstration (ACTD) study of several unmanned aircraft varieties in different performance classes. Among those varied types was Tier II+, focused on a nonstealthy, high-altitude reconnaissance UAV. The ACTD was planned and funded to satisfy a Defense Airborne Reconnaissance Office objective of fielding a capability for battlefield reconnaissance. With Maj. Gen. Ken

Israel, USAF, commanding that office and the U.S. Air Force designated as the sponsoring lead service for eventual transition to production, the most prized (and accordingly most emphasized) reconnaissance capability in Global Hawk prototype development was a primarily side-looking sector-constrained area search with a focused spot search inside that sector, and a further ability to cull out moving targets at the speed of a typical land vehicle. By definition, the robust reconnaissance sensor arrangement on Global Hawk was not optimally configured or coded for an all-aspect ocean surveillance capability that would be of interest to the Navy.

Despite an apparent mismatch in Global Hawk's intended mission for the Air Force and its potential repurposed mission for the Navy as a BAMS maritime patrol aircraft, the basic Global Hawk airframe and command and control schema would become the foundation for multiple Global Hawk derivatives for the Air Force and ultimately the Triton for the Navy. Triton's lineage can be traced directly to the Defense Department's ACTD Global Hawk prototype. Ultimately, though, by the time of Triton's approval for production, there remained many fewer apparent external similarities between Global Hawk and Triton, and fewer still internal similarities. In the course of Global Hawk's transition to a Navy-unique system, the agency governing the service's unmanned aircraft design standards was the Naval Air Systems Command (NAVAIR). NAVAIR's mandate for airframe structural enhancements, added-in capability for wing anti-icing, and upgraded anti-lightning protection would eventually differentiate Navy Triton from Air Force Global Hawk. Unlike Global Hawk, Triton's design is optimized to permit rapid descents from altitude below clouds to take still imagery or stream live video of surface targets, especially helpful for identification of unknown contacts that are not using transponders.

NAVAIR's rules governing Navy aircraft systems imposed significant enough change to the basic Air Force Global Hawk aircraft that the agency felt obligated to petition for a new alphanumeric designation; thus, it became known as the MQ-4C. Beyond the changes to the airframe, the MQ-4C also carries different mission equipment to optimize its effectiveness for maritime surveillance. The Triton, as the MQ-4C, is equipped with a newly developed multifunction active electronically scanned array radar designed specifically for spotting objects across swaths of seemingly endless tracts of ocean with 360-degree field of regard. This all-weather, all–compass-points-sweeping radar will detect ships and other objects of interest and take high-definition radar images, automatically cross-referencing contacts with an onboard automatic identification system or electronic support measures suite for classification or, in most cases, identification of contacts on the ocean surface. The Triton sensor suite also employs a multispectral electro-optical infrared camera at lower altitudes to track radar targets or stream live video to ground-based recipients.[7] For mandated due regard for other aircraft in international air space, Triton carries a sense-and-avoid radar and collision avoidance system that is, so far, a departure from similar capability envisioned by the Air Force for Global Hawk.

Yet much earlier in the life of Triton—even before it became known as BAMS—its creation was not countenanced by anyone at a senior rank in the Navy or an executive level in industry. When Northrop Grumman absorbed Ryan Aeronautical Company in the late 1990s, the company division that inherited the high-altitude UAV and was charged with carrying the DARPA ACTD through transition to engineering development and production was not staffed for, nor very mindful of, any Navy requirements that might dovetail with Air Force Global Hawk capabilities.

In a more tenured division of Northrop Grumman comprising the core remnants of the former Grumman front-end design staff in Bethpage, New York, interest in the maritime potential of the prototype was beginning to stir. Triton's operational capability today, as deployed with VUP-19, bears a remarkable similarity to the recommended adaptations originally set forth by that Bethpage design group's examination of Global Hawk almost two years before in the original BAMS concept study report in 2000. It will be seen in hindsight as an exceptional validation of and powerful testament to the vision, systems engineering ethos, trade study technique, modeling realism and fidelity, mission-mindedness, and generally close adherence to a governing naval orientation cultivated and groomed within that core Bethpage group over at least the prior twenty years—indeed, throughout most of the previous sixty years—of involvement in Navy aircraft design.

That small coterie of configurators and operations analysts elevated an Air Force–focused DARPA prototyping effort that was narrowly conceived and defined by the legacy Ryan Aeronautical Company's original work under the Tier II+ technology demonstration to give it new standing and a new mission role with a different service. To be sure, the Global Hawk UAV matured under the original Tier II+ contract was critical to the Bethpage group for providing a conceptual foundation upon which to build its study baseline, but the Triton that enters service with maritime patrol squadrons in the 2020s owes its genesis to the affinity for and understanding of the naval mission that still permeated the Grumman component of Northrop Grumman in the late 1990s.

Adapting the aircraft of one service to a different service's unique operating environment and mission demands a deep grounding in and intimate familiarity with both domains. The record is replete with successful examples of cross-service adaptation made possible by the original industry designer: Douglas Aircraft's redesign of its 1921 Navy torpedo-bomber as the Douglas World Cruiser for the Army Air Service's globe-circling endurance feat in 1923; Boeing's reconfiguration of its Navy F4B biplane fighter as the Army Air Corps' P-12 pursuit aircraft in the late 1920s; and, during the Vietnam era, Vought's A-7D reissue of its Navy-design Corsair for the Air Force and McDonnell Douglas's adaptation of the Navy F-4 Phantom II fighter aircraft for Air Force. These multiservice success stories in the tactical aircraft arena enjoyed greater success going from Navy to Air Force use than the reverse, due to the airframe requirements that must be designed in for carrier operations. These penalties for heavier structure to withstand the more taxing carrier environment are typically accumulated under the heading of structural carrier arrestment requirements, additive to the non-Navy variant, and are driven primarily by the requirement for a heavier centerline spar that can withstand the torsion and tension of catapult shots and arrestments.

No better example exists of this injunction against Air Force-to-Navy adaptation than the utter disaster that was the F-111B fighter-bomber aircraft of the mid-1960s. Even with Grumman's participation on the General Dynamics team to bring its well-established pedigree in carrier suitability design to navalize the Air Force version, the F-111A, no amount of redesign could effectively restore the structure for carrier suitability without imposing a crushing weight burden that made it incompatible for carrier employment. This effectively killed Navy involvement in what was termed the TFX project for its F-111B variant. More germane to this comparison may be the history of smaller drones that have been adapted for multiservice use, particularly by Northrop or its legacy company, Ryan: the BQM-34 target drone produced by Ryan, BQM-74 produced by Northrop, and BQM-145 from Ryan, all winged, jet-propelled vehicles the size of World War II German V-1 buzz bombs, intended to replicate airborne threats. Northrop Grumman and its legacy companies experienced some success in generating a stable, if only modestly profitable, drone business, not so much by a deliberate service specialization or tailoring as by building in features common to multiservice needs. That strategy proved viable through the Vietnam years and up to and well beyond Operation Desert Storm. But these systems, by their employment in several U.S. military services, became fixed on the capability spectrum as primarily target drones or more narrowly utilized as single-point reconnaissance assets or electronic signal emitters, typically regarded as expendable in both instances.

Drawing a closer bead on like examples beyond the tactical aircraft domain, the record is less ample. Considering land-based ISR as a point of comparison, the EC-121—the reconnaissance version of the Lockheed multi-engine Super Constellation passenger aircraft—stands out as one of the platforms common to both services: performing airborne early warning and fleet air reconnaissance for the Navy and the airborne early warning and control mission for the Air Force. The development of a large, robustly capable multiservice ISR aircraft, though, has been elusive at best, given the wide variance between the services' missions and operating environments. The EC-121 more easily conformed to the role of a Navy fleet air reconnaissance unit (a VQ squadron)—not unlike the same squadron employment in the Air Force. For the Navy, VQ never equated with VP (antisubmarine maritime patrol squadron). The mission and operating environment for VP diverges for the Navy with its need for patrol aircraft that not only perform visual and radar surveillance at altitude, but also perform well at near-wave-top heights to localize and prosecute submarine attacks.

Historically, in the antisubmarine role, the large patrol aircraft has had to exhibit an ability to operate effectively at low altitude through more radical maneuvering while dispensing sonobuoys and, in the end game, launching torpedoes. These two disparate operating realms have necessarily entailed some compromise in aircraft design, sensor configuration, payload bay, and expendable stores accommodation. In one sense, this design compromise is roughly analogous to structural penalties for carrier aircraft; here the penalty is assigned to high-altitude performance for having to carry accommodations for antisubmarine warfare functions. Yet maintaining two different aircraft separately optimized for the high- and low-altitude regimes could never have been justified inside the Navy's aircraft procurement and operations and maintenance budget authority. Furthermore, the manpower budget required to crew, train, and operate separate aircraft inventories has always militated against unique, single-mission configurations within the maritime patrol community.

Throughout the history of maritime patrol, there has not been a cost-effective solution to divorcing the two operating realms without carrying the burden of large crew, specialized sensors, and airframe-engine efficiencies from one domain to the other, along with the associated penalties. What resulted was very little trade space to adapt an Air Force aircraft—in this case, large aircraft hosting crewed positions for sensor operation for high-altitude surveillance—to a Navy requirement to perform part of its mission at high altitude and part of the same mission aggressively attacking submarine contacts at low altitude. Until the early 2000s no one had conceived a radically less costly way to do one part of the mission so that the other could be better optimized.

The different military services have different personalities, identities, and ways of viewing the world shaped by their history and experience in war. Those qualities color their analysis and interpretation of their own needs. Part of the problem is that differing service needs complicate the development process ... as the services attempt to address problems related to its suitability to their particular mission, costs inevitably rise.

—MATTHEW FAY, "The Problem with Joint Aircraft Development,"
Foreign Policy and Defense, January 27, 2015

What began as the high-altitude, long-endurance (HAE) UAV program and further evolved into the Global Hawk under the ACTD process was not, in its original conception, intended for any specific Navy mission or stated maritime requirement. The requirement that compelled the formation of a development program was a mission need statement, a first step in the formation of a military requirement. This was the precursor for eventually seeking resources with a budget line item and program infrastructure. The Joint Requirements Oversight Council, a senior Pentagon body of military and civilian officials convened to assess such military needs in the context of all other capabilities then in development, validated the mission need statement. This oversight council was, in effect, the guardian of all new start programs, assessing requirements for their legitimacy and ensuring that a duplicative effort was not already in progress. The mission need statement said, in wording de rigueur for this type of document,

> The warfighting Commanders-in-Chiefs [the regional four-star commands that were responsible of carrying out military action in their areas] have a need to provide commanders a responsive capability to conduct wide-area near-real-time reconnaissance, surveillance, and target acquisition, command and control, signals intelligence, electronic warfare, and special operations missions during peacetime and all levels of war against defended/ denied areas over extended periods of time. The evolution of the hostile surface-to-air and air-to-air threat and their collective effectiveness against manned aircraft and satellites can generate unacceptably high attrition rates. Current systems cannot perform these missions in a timely, responsive manner in an integrated hostile air defense environment without high risk to personnel and costly systems. There is a need for a capability that can be employed in areas where enemy air defenses have not been adequately suppressed, in heavily defended areas, in open ocean environments, and in contaminated environments.[8]

The phrase "open ocean environments" was a nod to the governing mandate of the time to provide the appearance of joint, interservice utility imposed on any requirement that expected to survive scrutiny by the oversight council. By a careful reading of the language of the mission need statement, this was, in reality, intended to satisfy an unmet need for overland reconnaissance and targeting of well-defended enemy ground forces. The Navy maritime patrol community had no role in formulating the requirement or staffing the oversight council outcome other than tacit signoff so as not to interfere with what was seen as an Air Force development effort that did not obstruct or duplicate any Navy effort. The Navy's single-minded interest in this mission area, as will be seen, was in engaging the acquisition process to produce, on a one-for-one basis, an updated replacement for the P-3C Orion large manned turboprop aircraft to fill out an inventory then nearing a threshold beyond which further structural life enhancements would not be cost-effective.

By their very definition, ACTDs were more about innovation than invention. Presumably the invention had taken place in a government, industry, or academic laboratory, and the underlying science was at least understood and probably had been demonstrated at a breadboard level that featured a live test in controlled laboratory environs. It had already exited the lab, having achieved that "eureka" moment of discovery. ACTDs provided the opportunity for an eventual military service user, with approval and oversight from the upper echelons of Defense Department management and active management by DARPA and, in the case of Tier II+, by General Israel's Defense Airborne Reconnaissance Office with Air Force involvement to define the employment and put the concept into play in a realistic operational environment. Subjected to the rigors of an operational environment, ideal laboratory conditions were suspended, and the potential military user was better able to assess its usefulness in near-real-world settings and to determine whether the weight of favorable utility measures supported transitioning the concept to serial production. The innovation was in determining the prototype features that could best be measured and operational employment that could best stress the prototype to ascertain its potential military utility.

To be sure, Global Hawk represented an innovative new concept that was an amalgam of new technologies and aerodynamic design. Ryan Aeronautical Company "had experience building UAVs similar to the Tier II+, so there was some engineering trade space to borrow from their past projects to make the cost work while still meeting the schedule and most of the performance objectives."[9] But the real innovation was in finding Global Hawk's operational niche where it might transform ISR by broadening and extending the mission for longer dwell and endurance and employing fewer operators for lowered cost, done within the rhythm and practice of Air Force ISR operations. With a successful demonstration, the potential existed to favorably alter ISR force structure and, with it, long-range investment decisions about force modernization and recapitalization for the Air Force.

In that Ryan succeeded in capturing the DARPA technology demonstration, and Ryan Aeronautical's new master, Northrop Grumman, succeeded in brokering the cross-service adaptation to fit a Navy need that was not immediately apparent to the Navy maritime patrol and reconnaissance community, it had surmounted steep climbs that required well-reasoned and executed capture strategies. That Northrop Grumman then husbanded the development through yet another industry-wide open competition to its full realized potential was an exceptional achievement—not a foregone outcome by any means. Triton's fifteen-plus-year march from concept to initial operating capability—with its desired capability fully resourced and ready for worldwide deployment—was fraught with setbacks, miscues, and fits of indecisiveness compounded by misperception and partisanship on one hand, and examples of sterling leadership by both government and industry officials on the other. At key moments, exceptionally strong leadership permitted—actually encouraged—small teams of individuals to form, as much with mixed Navy and industry membership as with all-uniformed Navy groups or all-industry participants, to break with convention and run against the entrenched attitude or culture. Those small teams, by happenstance, were composed of exactly the right balance of experience, outlook, and determination to fully exploit the space the leaders defined for them. At each turn, the emergence of leaders, exhibiting very definable and empowering leadership styles, made the difference by granting license to the breakout innovative thinking needed to drive a concept through opposition or indifference to the next rung.

As much as it is a chronological narrative of Triton's success, of the creativity and persistence of those small teams that were key to that success, this telling is an exposition of leadership styles that worked—shaped by the prevailing service or company culture or bearing influence on that culture—and those leadership styles that proved wholly ill-suited to the situation at hand and nearly derailed Triton in the early BAMS stages. It is told against a backdrop of institutional cultures that were in constant flux, which informed strategy at each of several stages but would also complicate and stymy the execution of the strategy. Fittingly, the first two chapters set up the narrative, provide context for the chronology that follows, and present leadership definitions useful to gauging the performance of key individuals identified in the story. In the ensuing chapters, the story follows a timeline that begins with the Ryan Aeronautical Company's capture of the DARPA technology demonstration and the birth of Global Hawk, continuing through the internecine struggle within Northrop Grumman to settle on the shaping strategy, the effort to transition Global Hawk capabilities to Navy performance in the maritime patrol and reconnaissance mission, and culminating with the corporate-level strategy to compete and win BAMS in a fiercely competitive environment among industry titans and to preserve that win when protested by the losers.

Chapter 3 establishes the set of events wherein the opportunity space was created by intuitive and forceful leaders at DARPA, was quickly filled by bright, adaptive engineers at Ryan Aeronautical, and was embraced by mid-grade early adopters in the Air Force. Here, the vision and organizational skill of one leader at Ryan, Bob Mitchell, drove the efforts of his company and broke through the industry farrago of competing UAV concepts, revealing the keen discernment and business judgment that propelled Ryan into the winner's circle with its Global Hawk UAV.

Chapter 4 maps the larger geopolitical and force structure movements that opened an opportunity space in the Navy for reinvigorating a warfighting community's culture by altering the scope of its operational concepts. A visionary disrupter in uniform in the person of Capt. Alan Easterling broke with the longstanding orthodoxy of the manned aircraft maritime patrol and reconnaissance community to embrace a more affordable hybrid solution to modernization that rested on incorporation of UAVs and ultimately showed the community a way to achieve recapitalization and more effective warfighting capability. He was matched by an industry disrupter in the person of Howie Frauenberger, who artfully conceived of the hybrid solution and made a partnership with Easterling to upend the status quo.

Chapter 5 details how the concept for a Navy HAE UAV was initiated and coalesced around a radically new operating logic for maritime patrol and reconnaissance. Next, the radically new concept of a hybrid mission, proffered by Frauenberger and embraced by Easterling, became the basis of a comprehensive socialization campaign to change minds, formed and executed by Maris Lapins and Ernie Snowden.

Chapter 6 shows how, as the shaping campaign built a head of steam, a growing base of advocates help propel the concept forward, just as competitive pressures forced the BAMS team in a new direction. The authors offer observations here on best practices for moving an acquisition program forward in its early conceptual stages.

Chapter 7 explains the pause in the expected progression of BAMS development due to cultural and bureaucratic opposition that threatened to derail BAMS but was ultimately mitigated by analysis and fleet demonstration. Leadership failures in the Navy and more so in industry hamstrung further progress of the UAV concept. Only when the disrupters, Frauenberger and Easterling, reengaged was the earlier course resumed with the initiation of the persistent unmanned maritime airborne surveillance study that would underwrite the competitive win against industry titans.

Chapter 8 offers a view into the ultimate resolution of the drama. A dynamic leadership style again emerged, creating the conditions for a new industry and Navy team to drive a refreshed strategy to conclusion, fulfilling the promise of the first act but for the Navy. Early visionary Bob Mitchell returned to reset the strategy under the most pressing competitive conditions. A new strategic leader, Bob Wood, entered the story to partner with Mitchell to focus and execute the strategy and to effectively serve as alter ego to Bob Mitchell on the winning proposal team.

Chapter 9 reaffirms the strength of capture leadership and strategy in fending off serious challenges to the prize now won, challenges that could have potentially set back the successful introduction of unmanned systems to regular Navy service. Finally, chapter 10 shows how the vision and foresight of Mitchell and Wood set in place the defensive strategy that allowed the winning BAMS concept to retain its prize in the face of determined protest.

As much as the high-altitude long-endurance UAV ACTD was about innovation in Air Force ISR operations, the concept study phase and experimentation that followed in the Navy were about re-innovating the concept to introduce the technology in an entirely different context: maritime patrol operations reimagined in a way that added value by infusing into the mission calculus a large-scale UAV with previously unattainable range and endurance. Importantly to its eventual success, the UAV had to be positioned in a way that did not distract from the maritime patrol community's single-minded pursuit of its manned airplane replacement. If the UAV could be presented as a means to achieve recapitalization of the manned aircraft, it would earn its way. It depended, of course, on whether the maritime patrol community could be persuaded that the odds of obtaining a new manned aircraft in a constrained fiscal environment improved substantially with the introduction of an unmanned element into the mission—that is, improved to the point of resurrecting the community from its then-ongoing track into obsolescence and decline.

To better grasp the nature and difficulty of the cultural gap that had to be bridged to move the Global Hawk from the Air Force into serious consideration by the Navy, it is instructive to more deeply examine the patterns of Navy institutional thinking about innovation. The work of two prominent academic theorists, Dr. Barry Posen, director of the Massachusetts Institute of Technology's Security Studies Program, and Dr. Vincent Davis, former faculty member at the Naval War College and researcher at the Center for Strategic and International Studies, bears particular relevance. Both started from a view that a commonly accepted definition of military innovation pertained: innovation is the melding of two components, the technical and the doctrinal, with doctrine denoting a comprehensive and practical understanding of the use or employment of the technological.

Posen examined military innovation among major European powers, including the Royal Air Force and its aircraft development, in the interwar period. His study led him to conclude that doctrinal innovation more often is imposed from outside the military through the intervention of civilian authorities and that technological innovation originates in the persistent advocacy of unconventional, even nonconformist mid-grade military officers. For Posen, "Military culture in general, and naval culture in specific, has good and historically documented reason to hedge against risk inherent in fixes that rely on fundamental departures form the normal way of doing things."[10] We see in Posen's work an acknowledgment of an institutional inertia, particularly in naval communities, that militates against deviations from commonly accepted employment schemes and tactics. Davis, on the other hand, assessed case studies that detailed the Navy's efforts to develop a carrier-based nuclear delivery capability and the development of the Navy's ballistic missiles. He concluded that technological innovations most often "reflect the belief that innovation represents a better way of doing a mission already inherent in the Navy . . . and comes predominately [*sic*] from the personal initiative of middle grade officers (O-4 to O-6). . . . these innovators generally recognize rather than invent, possess unique technical expertise in the area of innovation, and form horizontal working alliances to promote the innovation."[11]

Posen's and Davis' work was surveyed by Navy Capt. Bradd Hayes and Douglas Smith in "The Politics of Naval Innovation" and examined for applicability against other test cases, mainly cruise missile and Aegis development. Hayes and Smith drew out one more strand of this characteristic cultural resistance to innovation: the "tendency of the Services to reject applications not under their control. This aspect of institutional behavior is highlighted in the Air Force recalcitrance at accepting the Office of the Secretary of Defense (OSD) designation of the Navy as the Executive Agent for development of all variants of the cruise missile."[12] Hayes' and Smith's conclusions are especially relevant to the reticence shown by the Navy toward an Air Force–sponsored high-altitude long-endurance UAV ACTD for which a Navy mission had not been yet directly demonstrated. Posen's and Davis' work, and Hayes' and Smith's analysis of it, bears directly on

The essence of just about everyone's definition of a military-technical revolution points to the coming together of two components: technology and doctrine. It is not just technology. That's the reason it's called "technical" and not "technological." It's the marriage of technology and doctrine, with doctrine representing a collective understanding of employment.

—SAM GARDINER, "The Military-Technical Revolution,"
RSAS Newsletter 4, no. 3, August 1992

the experience of those attempting cross-service adoption of Global Hawk. Their experience illuminates the cultural realities that, in turn, heavily influenced the strategy and tactics waged against service bias. Hayes assessed that "the Navy, like most large bureaucracies, resists change, innovation does occur, and no single source for it can be identified."[13]

At the time of Hayes' and Smith's study, the catchphrase entering the lexicon throughout the Department of Defense was "revolution in military affairs." In a retrospective critique, an early avatar of this revolution among Pentagon strategists, Dr. Andrew Krepinevich, captured its essence: "A military-technical revolution implies that new technologies are not simply used to conduct military operations more effectively, but that they are used to conduct more effective kinds of military operations. In addition to the intellectual task of identifying and understanding new ways of warfare, there is the practical task of getting military organizations to adopt those new ways, and even to adapt themselves to those new ways. There may be special problems with innovation in this military-technical revolution since much of the innovation may have to cut across Service boundaries."[14]

Taken together, these studies point to a prevailing cultural bias that, as a first order, animated the maritime patrol community to oppose any dilution of their well-honed and comfortable operational practices; second, it hardened resistance to any deviation from Navy control over technological development or—especially—change in doctrine that implied Air Force influence and control; that the bias could likely only be overcome by outside nonmilitary agents with inventive and compelling new logic; and finally, that the bias would require reinforcement by uniformed mavericks in the community pushing internally for innovation.

CHAPTER 2

DISRUPTERS TO THE FORE

Disruptors don't have to discover something new; they just have to discover a practical use for new discoveries.

—JAY SAMIT[1]

Triton is a reality today primarily through the early efforts—and, to a great extent, those in the final phase of competitive acquisition—of Northrop Grumman acting as the outside agent of change, and the dogged, near-fanatical persistence of key mid-grade and senior officers in the Navy program management office and on the staff of the Chief of Naval Operations (CNO) who grasped the vision—who could perceive a new, more cost-effective way of doing a familiar mission while also recapitalizing equipment for that mission. On the CNO's staff, this zealotry was more tempered, as officers had to daily conform to the rhythms of the planning and programming process, where their attention was fully captured by the constant resource tradeoffs that remain the daily diet of the CNO's staff.

Operating at the other end of the scale, the prevailing Pentagon "transformation" mantra provided more room in the acquisition environment for entertaining new inventions and fresh technological perspectives and for engaging any number of breakthrough concepts presented by government labs or industry. These two distinct subcultures—embodied in resources and acquisition—tended to foster different models for success and different behaviors to fit the respective environment. The behavioral component to these subcultures manifested in the way officers in each approached their tasks and in the leadership styles that emerged and held sway. In the resources case, so much of the routine was driven by process mandated by statutory requirements and Department of Defense (DOD) instructions and guidance that leaders functioned most effectively by simply setting a vision and purpose for subordinates and checking progress by measuring success against process milestones—with little impact on cultural change.

Throughout this narrative, leaders that fit this type are referred to as "traditionalists." The traditionalist has succeeded inside the bureaucracy by learning and applying process to achieve success. He or she is perceived by followers as having an unimpeachable resume and a string of professional accomplishments that, by their association with and allegiance to the traditionalist, will convey to them. A necessary ingredient for that association as well is the perception that the traditionalist enjoys the standing and high regard of his or her superiors and will willingly stand between followers and superiors to praise or protect the subordinates with those superiors.

Conversely, officers who are more iconoclastic and willing to deviate from accepted practices by defying convention are termed "disrupters." They typically possess a well-developed self-confidence, are quick to grasp the longer term or more altruistic view, and are much more willing to entertain innovation, to challenge bureaucracy, and to shake up the status quo to obtain results that serve the long view. Their followers are typically of similar ilk and often feel that by their association with the disrupter, they are joining an embattled movement for the better outcome.

Cdr. Russell Schuhart draws a similar analogy to these traditionalist and disrupter labels, terming them "company men" and "free agents." In his apt ordering of these two types, he notes that the company man may "bounce from assignment to assignment attempting not to disturb his surroundings" and may actually seek to avoid those situations, usually occurring as they rise to higher rank, that require making significant and career-impacting decisions. He describes free agents as "those who think for themselves, make decisions based on the particulars of the situation," and are prompted to make decisions regardless of the perceived negative consequences to them personally: "Company men will rarely innovate, out of fear of failure, but free agents constantly innovate, seeing failure as a path to improvement."[2]

These absolute characterizations may be more nuanced in real, every-day people. Officers are most often presented opportunities for action that come wrapped in a host of situational factors that tend to blur the lines for making clear, decisive judgments. As will be shown, both types may be found in any of the domains that encompass acquisition, requirements, or operations. None of these domains are immune to successive layers of bureaucracy and the rule set for every participant that is constrained by process. The disrupter (or free agent, in Commander Schuhart's writing) can find equal opportunity to stretch the process boundary in any of these domains, but as a generalization, acquisition and operations will be more attuned to accommodating creativity and innovation. Disrupters bring an evident sense of determination that can often propel breakthrough success by nurturing new technologies or concepts and advocating for the unconventional when they can make a strongly argued case. They are often most successful in transmitting cultural change when, through direct, hands-on involvement, they put a considerable professional repu-tation at stake to persuade and enlist advocates throughout the bureau-cracy across community boundaries.

Arguably, though, there is a third model of behavior exemplified by the "inspirational" leader who may rise from either the traditionalist or disrupter example but who employs a measure of both to galvanize and incite followers to greater purpose and accomplishment. Robert Herbold, former chief operating officer of Microsoft, explained it as "managing people's hearts and minds . . . from the necessity of making tough choices and the importance of strong vision for the future to creating a culture of curiosity and innovation."[3]

Closer to our narrative in time, two officers arrived at their respec-tive assignments on the CNO's staff and in the maritime patrol program office—one in December 1999 and the other a little over a year later—and, by dint of their personalities and behaviors, exemplified a blending of leadership styles. The two—Rear Adm. Tom Kilcline and Cdr. Alan East-erling—arrived at a critical moment in the inception and gestation of what would eventually become Triton but that was still forming as BAMS. Both officers were graduates of the Naval Academy, and they were compiling admirable service records in their respective naval aviation communities, fighter and maritime patrol. Both were steeped in family pedigrees notable for long and distinguished service to naval aviation. By a remarkable confluence of timing and circumstance in the early 1980s, their fathers, both vice admirals, were concurrently serving as commander, Naval Air Forces, U.S. Pacific Fleet and commander, Naval Air Forces, U.S. Atlantic Fleet. Neither of the two officers made much of this coincidence.

In 2001 Kilcline reported to director of air warfare Rear Adm. Mike McCabe on the CNO's staff and to begin his tenure there was assigned the lead for a newly established division designated "X." His role was to bring order to the incoherence that was the planning and programming process for a disparate set of unmanned vehicles, then just beginning to gain some visibility in wider naval aviation circles. This duty presented a steep but manageable learning curve for an aviator whose flying experience was in the realm of tactical manned aircraft flying from aircraft carriers. The grouping of UAV requirements officers in one Pentagon office, partitioned from the manned platform requirements officers inside the colloquially named Ranch (where ranch hands herded operational and budget needs for mainstream or primary Navy aircraft) tended to blur any readily discernable connection of a UAV to any particular warfare mission, and thus tended to make orphans of unmanned systems. Predating Kilcline's arrival, this set-up created built-in hurdles for gaining much traction for unmanned systems by denying them the sponsorship and advocacy of the primary manned warfighting platforms in the recognized mission areas.

To make the most of a difficult organizational construct, Kilcline adopted a leadership style best described as inspirational that set boundaries for his "X-men" and motivated them by extolling the important role they would be playing in introducing a new class of aircraft to naval aviation. Kilcline was very much an inspirational leader type in the sense that he laid out a broad vision for where the Navy logically should go with UAVs, and he compelled his group to conform to the annual rhythm and pattern of the planning and programming process. In so doing, he set an important precedent and enduring expectation that unmanned systems would hold a regular and predictable place in the buildup of the budget request. At that moment, it was beyond his station and pay grade to greatly influence how the UAVs fared collectively in the prioritization of air programs for obtaining an equitable share of naval aviation resources in the three-star executive board that decided Navy-wide priorities. He relied heavily on his individual staff members to provide the mission context for the respective UAVs to which they were assigned.

Lt. Cdr. Mark Turner, possessing a significant maritime patrol background, was drawn in from the maritime patrol shop in the Ranch to fulfill the role of requirements officer in the X shop for what was emerging as a preeminent UAV issue, by then known as BAMS. With top cover and minimally intrusive guidance from Kilcline, Turner seized the opportunity to stake out his own preserve as the recognized face of BAMS and the chief CNO staff champion in the next round of issue paper development from which sprang platform funding lines. Under Kilcline, Turner was authorized to exercise a vigorous advocacy throughout the program proposal cycle, abetted by his already established good standing with the maritime patrol shop in the Ranch. As the cycle advanced to higher levels of review, he persisted in making the successful arguments to one-star and two-star boards. Finally, he was able to run counter to conventional wisdom by securing a steadily increasing funding wedge for BAMS in the budget request buildup.

Turner's success, ahead of a staffed and formally approved requirement and without the benefit that such approval conferred as an authorized new start, was a testament to his quick mastery of the process and his persistence once engaged. That Turner made considerable headway in arguing for that maritime patrol mission linkage for BAMS inside the process is evidenced by officer fitness reports from Kilcline that characterized his performance in uncontestable phrasing: "unrivaled professional acumen; expertly guided the Broad Area Maritime Surveillance (BAMS) UAV effort, one of the most visible programs within DoD.... His requirements and acquisition insight took the BAMS UAV program from zero to $2 billion budgeted funding within an eight month timeframe."[4]

But before Rear Admiral Kilcline and his requirements officer, Lieutenant Commander Turner, had their turn with BAMS and could map a funding stream with accompanying logic for building out an unmanned inventory, the all-important connecting sinew with maritime patrol had to be conceived, analytically and fiscally justified, and introduced in a compelling way to a community that still saw itself as airborne warriors in a manned aircraft community. That task fell to Easterling, though upon his arrival in the program office for maritime patrol in 1999, an unmanned aircraft fulfilling an adjunct role to a manned maritime patrol aircraft was not in his or anyone's scan (except possibly the small cell of visionaries inside Northrop Grumman's Bethpage operation).

In an office absorbed with upgrades and sustainment of the manned P-3C Orion turboprop aircraft inventory, Easterling's singular focus as the deputy program manager was initially on the recapitalization issue: the manned aircraft replacement for the P-3C that had been labeled as the multimission aircraft. Kilcline was a traditionalist, and his counterpart in the traditionalist mold on the acquisition side was probably the maritime patrol program manager to whom Easterling reported as deputy, Capt. George Hill. He was nearing the end of a career of honorable service and would probably retire if not selected for flag officer. He was not inclined to play the part of firebrand in the interim. Easterling, however, was the disrupter, an insurrectionist who, despite a conventional career track to this point in the maritime patrol community, possessed the sort of intellectual curiosity that allowed him to envision an alternate future for the mission that he knew well. Still committed to the MMA, Easterling's studied understanding of Navy and defense budget priorities in the immediate post-Soviet era (much of the will and ability to carry on submarine operations dissolved along with the Soviet Union) prepared him for easier acceptance of nonlinear solutions to innovative concepts. Events of the next few years showed him to be by nature more receptive to unconventional notions, even radical concepts that could lower the MMA development and procurement bill and/or alter the mission doctrine and tactics in a way that ushered in a new, more cost-effective technological approach.

At least a half-dozen years before Kilcline, Turner, or Easterling had countenanced any notion of an unmanned aircraft and the role it might play in the maritime patrol mission, other events were unfolding—bigger movements in the global geopolitical balance, the CNO's staff and naval aviation acquisition organization, defense procurement practice, and defense industry fortunes—that would converge in time to create the conditions for a successful BAMS introduction. Kilcline, Turner, and Easterling arrived at their new assignments at the end of the 1990s, a decade of precipitous decline in overall Navy budget and of significant drawdown in Navy force structure. A fundamental shift in the global power balance and a reordering of priorities in response had the effect of altering the staff relationships and processes they would employ, if not the very environments in which they found themselves working.

At the start of the 1990s, a new CNO, Adm. Frank Kelso, wrought profound changes to his staff and to the planning objective memorandum (POM), part of the programming aspect of the planning, programming, budgeting, and execution process. The assistant CNO platform sponsors (colloquially referred to as the "barons") were downgraded from three-star to two-star positions and reported to a three-star deputy CNO performing the "requirements and resources integrating" function. The immediate and lasting effect was to diminish the power of the aviation baron to inveigh against and cajole his counterparts in the other warfighting communities with well-crafted arguments before the CNO. The historic prefix for CNO staff positions was changed from "OP" to "N" in recognition of the staff-code convention prevalent in the U.S. Army and Joint Staff. That change in title brought with it more deep-reaching change in process, especially for integrating joint service considerations into the Navy POM build. Kelso drove the staff interplay away from the "perceived traditional model of competition (among barons), with expert analyst intervention, and imposed opaque high-level decision to a new model: junior flag officer open dialogue, consensus-building and agreement."[5]

A significant part of the change that would persist beyond the decade and much later influence the Navy and the industry BAMS strategy was a renewed emphasis on wargaming, mission analysis, and modeling and simulation, best expressed by the embrace of joint mission area assessments as a foundation of the POM buildup. For the balance of the decade, two more CNOs enhanced but did not deviate greatly from Kelso's changes. Nearing the end of the decade, under Adm. Jay Johnson, several refinements would bear directly on the success of the BAMS strategy. Kelso's mission area assessments were superseded by Johnson's more

collaborative process, which engaged integrated teams of fleet operators, system command representatives, and warfighting requirements
specialists as well as emphasizing cost savings obtainable from new programs and offsets that could be derived from modifications or outright
cancellations.

Perhaps the most tangential but impactful changes instituted by
Admiral Johnson at the end of the 1990s—changes that would bear on
the formulation and successful execution of the Navy's BAMS acquisition strategy and Northrop Grumman's BAMS capture strategy—were
the creation of the Navy Warfare Development Command's maritime
battle center as the sponsor and organizer of fleet battle experiments—
laboratories for innovative warfighting systems and techniques—and
the selection of Vice Adm. Art Cebrowski as the three-star president of
the Naval War College. Cebrowski tirelessly used the college, the intellectual and spiritual center of the service, as a platform to proselytize
for the concept that would become known as network-centric warfare,
progenitor of FORCEnet.[6] FORCEnet was intended to form the back-
bone for command and control of disparate elements of naval and joint
forces by exploiting the power of networks. In the decade that followed,
FORCEnet—the concept—would be exhaustively amplified by engineering investigation and play in fleet battle experiments. As a potential node
in the FORCEnet universe, BAMS would eventually become, by design,
an enabling element of the network as overlaid on maritime patrol.

With Admiral Kelso's moves to establish a more integrated and collaborative team approach to the POM build, the arrival of Vice Adm. Bill
Bowes as the commander of the Naval Air Systems Command signaled
a shakeup in the organizational approach to staffing for aircraft acquisition. Mirroring the CNO staff realignment to integrated product teams,
Bowes established a competency-aligned organization. In execution, this
entailed the breakup of longstanding "homesteads" in stovepiped major
program offices, creating in their place a program staff matrix of functional experts drawn for defined periods from a competency homeroom.
This spelled a diminution of power previously consolidated in a single
aircraft program manager, requiring a more deliberate and thoughtful
collaboration between program manager, program executive officer,

and competency leader to agree on a plan for manpower and function expertise. An expected benefit was the assignment of exceptional engineers, logisticians, and business managers to emerging higher priority programs on an as-needed basis. Some long-time NAVAIR engineers and analysts were wary of the change at the time and were even disdainful of its alleged potential for greater efficiencies. Among them was a true NAVAIR institution in the person of George Spangenberg, evaluator on the F-14 and a number of significant aircraft programs going back to the 1940s.

In his oral history, Spangenberg dismissed integrated product teams as a nod to popular management buzz words, deriding their promise of improved management efficiency. When Bowes's successor, Vice Adm. John Lockard, dutifully implemented a relocation of the NAVAIR headquarters from Arlington, Virginia, to Naval Air Station Patuxent River, Maryland, to comply with a congressional base realignment and closure edict, Spangenberg had more choice words: "I cannot conceive of a move of this nature being countenanced by any of the flag officers under whom I served. Distancing the technical community from OPNAV and the Congress is, in my opinion, a grave mistake. All the advantages we in naval aviation's technical community had over our Air Force counterparts in Dayton have been ignored. Even more serious, perhaps, is the precipitous loss of experience occasioned by earlier than planned retirements and resignations of those unwilling to move from the DC area to southern Maryland."[7]

Spangenberg's critique notwithstanding, the principal benefit to the maritime patrol program management office and the evolution of BAMS was twofold: from inside the rebuilding engineering competency at the new Patuxent River location, several truly competent and exceptionally devoted Navy civilian engineers were beneficially detailed to the emergent BAMS program; and, wherever a shortage of civil service engineering talent threatened the initial configuration study, the program management office proved adept at bringing on and using competent industry engineering capability. In the earliest days of the BAMS concept, the application of Northrop Grumman's advanced concept work was particularly helpful with operations analysis and design trades until the Navy

Department's bureaucratic impulse for competition at all costs kicked in and severed much of that relationship. Nonetheless, these factors, closely associated with the headquarters relocation if not directly influenced by it, proved fortuitous, in the first instance, in bringing in Craig Nickol as deputy to Easterling and, later, Joe Landfield as chief engineer; and, in the second instance, giving Northrop Grumman a preeminent seat at the table, in recognition of its incumbency on the DARPA high-altitude long-endurance UAV project, that provided a path for shaping a Global Hawk–centric requirement as most appropriate for BAMS.

In several respects, 1993 proved to be the seminal year in the genesis of BAMS, an inflection point at which a newly elected Democratic administration replaced an outgoing Republican administration with its own worldview and defense priorities. Turning abroad, the new administration confronted a much-diminished Soviet Union and the tantalizing prospect of a peace dividend that could be recouped immediately. To be sure, the stunning victory of U.S. and allied forces in Operation Desert Storm just two years earlier had a collateral effect on U.S. military thinking and on the new civilian leadership's view more generally about future needs for ISR in a reordered world. Much of the equipment used in Desert Storm validated its usefulness against an armor-heavy land force that approximated the former Soviet Union. But with the Soviet Union now depleted, many believed that the Desert Storm force was wrongly configured for potential regional contingencies that appeared on the horizon.

The Pentagon's cancellation of the Quartz program in 1993 was a first indicator of the new thinking taking hold in future force plans, and it set in motion a long train of program decisions influenced by transformative acquisition practices and budget realities, ultimately contributing to conditions favorable for the appearance of BAMS. Cloaked in secrecy from its conception in the early years of the Ronald Reagan administration, Quartz was a robust unmanned reconnaissance system intended to "fill a gap between episodic satellite passes and limited manned reconnaissance flights by not only finding [mobile intercontinental ballistic missiles], but loitering for long periods of time to track their movements."[8] Often referred to in media accounts as Aurora, Quartz challenged aeronautical designs of its day by its requirement for transcontinental range and extremely high altitude. In the aftermath of Desert Storm, and in light of new OSD guidance to focus on lesser regional contingencies, Air Force interest in Quartz waned. When the development schedule flagged and program costs escalated exponentially, the Air Force walked away from Quartz.

That year, in congressional testimony, commander of U.S. forces in the Pacific, Adm. Richard Macke, spoke on intelligence needs in his area of responsibility: "We need the data that a tactical, an SR-71, a U-2, or an unmanned vehicle of some sort will give us, in addition to, not in replacement of the ability of satellites to go around and check not only on that spot, but a lot of other spots around the world for us."[9] In that same year, the Joint Requirements Oversight Council, the most senior military and civilian review board consisting of the service vice chiefs and chaired by the vice chairman of the joint chiefs, recognized a compelling need and validated a proposed acquisition path for a three-tier system to nurture UAV programs toward fulfilling the requirement described by Admiral Macke. The first tier included low-altitude, less robust UAVs; the second tier consolidated medium-altitude, long-endurance UAVS, of which Predator was the most publicly visible example, and a high-altitude, long-endurance UAV, no outstanding examples of which were then in the inventory or in development.

A further requirement to exhibit stealthy survivability characteristics potentially complicated any future designs and would require careful management. Of underlying importance to the foothold gained by the high-altitude, long-endurance UAV in the Navy inventory, the vice chairman of the joint chiefs who guided this oversight council outcome was Adm. David Jeremiah, who in retirement would resurface as a behind-the-scenes advisor to Northrop Grumman's BAMS capture strategy. But at this moment, because of the perceived dual and conflicting requirements for high-altitude endurance and survivable penetration at lower altitude, the stealthier concept generated warning flares for incoming defense officials with the new administration.

Unlike prior intelligence-gathering systems that enjoyed a veil of secrecy (and, behind that barrier, near unconstrained funding), the stealthy concept would be pursued more openly. With that openness, these new defense officials were, if anything, exceedingly mindful of cost containment in the era of the peace dividend. A stealthy UAV presented a risky option for cost control going forward, and it invited concern. Pentagon planners wisely decided to bifurcate the concept into two separate pursuits: Tier II+ and Tier III–, wherein design objectives could be more narrowly focused and cost, theoretically at least, more carefully controlled. Tier III– became the survivable penetrating intelligence platform, and Tier II+ would be the HAE UAV. In the summer of 1993 Tier II+ and Tier III– gained new impetus from a Defense Science Board (DSB) study that "called on the Pentagon to spur development of UAVs. This, it said, would help 'fix the problems exposed in Desert Storm.' The Defense Science Board said the use of reconnaissance UAVs would help U.S. forces gain wide-area coverage, acquire all-weather access to the battlespace, and integrate combat information. The challenge was to create a new type of craft, one that would build on the experience gained from decades of operating remotely piloted vehicles—or drones—and experimental high-altitude vehicles."[10]

Witness testimony before the Defense Science Board by intelligence officers with Desert Storm experience cited a desperate need to photograph large swaths of contested geography with minute resolution on the order of one foot or less. The requirement that they described and that the DSB duly recorded in its study was for persistent uninterrupted coverage of an area the size of the state of New York over a twenty-four-hour period. In terminology picked up by the Defense Science Board, "broad area" became the most apt descriptor of the requirement, would take hold in the vernacular of the intelligence community, and would stick even as the requirement later expanded to include maritime surveillance. The DSB further determined and recommended that the most efficient means to acquire this broad area surveillance capability would be an acquisition strategy that relied on commercial off-the-shelf products to the greatest degree possible, gained by a hybrid or abbreviated "special" acquisition protocol.[11]

By the end of the 1990s, the stage was nearly set for a Navy BAMS strategy to be put into motion. But to reach that point, events coming into sharper focus in 1993 had to play out through the years that followed.

A new capability had to be incubated under special acquisition guidelines that shielded it from the usual pressures of weak service advocacy, low prioritization in the resourcing tradeoffs, and obfuscating layers of bureaucratic scrutiny that short-circuited many other programs in the mainstream of defense acquisition. It was not at all clear in 1993 that the Air Force, as an institution wedded to manned aircraft as the core logic of its force structure and its operational employment, could evoke the advocacy and energy required to champion a new priority for an unmanned aircraft—especially in a mission area where a user community with congressional and industry supporters already existed: the U-2. The Air Force was not alone in this. The Navy maritime patrol community, hit hard in the early 1990s by steep force structure drawdowns, was desperately trying to expand its presence into other missions to preserve officer and enlisted billets and also to give its community new purpose in a world defined by lesser regional contingencies, brushfire conflicts over undeveloped terrain, peacekeeping along unstable borders, and coalition-building for unfettered access to the world's littorals.

The 1993 DSB study urging more investment in UAVs reflected a much broader push across the landscape of military unmanned systems. That year, Grumman engineer Howie Frauenberger found himself assigned to lead design trade studies on a potential U.S. Army command and control airborne vehicle. Seen by the Army as a small rail-launched, parasail-recovered unmanned aircraft for battlefield intelligence and reconnaissance, Grumman's land combat office in Great River, New York, was the business unit responding to the Army's solicitation. Frauenberger came aboard midway into the proposal preparation only to find that the candidate aircraft had already been chosen, a UAV designed and built by a small company in Georgia that made fiberglass bathtub enclosures. The Great River business unit's director had made the choice, even though he had no aeronautical design experience.

Frauenberger did an exhaustive analysis of the chosen system—a delta-wing design—against a notional proxy design with a more conventional straight wing. When called before the business unit's senior management for a proposal review, he was told to brief the official position that the chosen UAV as it existed met all of the Army's specifications.

Frauenberger briefed it straight from the shoulder, with no varnishing of his analysis: the chosen system was a loser, but it could be salvaged with a relatively easy wing reshaping. Frauenberger was dressed down by the unit director after the meeting and received his layoff notice two weeks later—only to be saved by senior leaders in the aircraft side of the company at Bethpage.

This was the nature of Frauenberger's persona: a stubborn confidence in his own aeronautical design expertise and the assertiveness to resist any persuasion otherwise. That design expertise was substantial; it began thirty years earlier as a new Pratt Institute graduate hired by Grumman as a thermodynamics engineer on the Apollo lunar module. His next two decades of work included development of advanced fly-by-wire flight control systems, four years as staff assistant to Grumman's chief technical officer, and two years seconded to the Grumman vice president of engineering as his troubleshooter on aircraft acceptance by a varied list of U.S. and international customers. He topped off this impressive experience as a program manager for design, fabrication, and testing of a National Aeronautics and Space Administration–managed space shuttle–attached on-orbit experiment supporting the Strategic Defense Initiative. It was here, more than in all the previous work, where Frauenberger came to realize the extremely positive value of cultivating and sustaining a great rapport with his customer by means of close collaboration on joint customer and contractor teams and by ready responsiveness to customer concerns during contract execution. This view would be a characteristic of his most important work to come: as an innovator of unmanned concepts of operation for a different customer—the U.S. Navy—and a disrupter of the staid culture that held the maritime patrol community in a firm grip.

CHAPTER 3
TURBULENCE

Washington Post–ABC News nationwide random poll in April 1993 pointed emphatically to deep-seated public unease about the level of federal spending. Fully 73 percent of those surveyed after President Bill Clinton's first one hundred days believed that his administration was not moving fast enough to bring spending down.[2] Defense spending was overdue for steep reductions, which would be a focus of the new administration. New DOD leadership pledged to drive defense spending further into conformance with the expectations of a sizable peace dividend. The fast-withering Soviet Union was, by 1993, judged to be a much-diminished threat to national security, lacking the resources and cohesion, if not intent, to seriously challenge the U.S. military. Secretary of Defense Les Aspin, having served the last eight years of his sixteen terms in Congress as the chairman of the House Armed Services Committee, brought to bear a distinctly contrarian perspective to business as usual inside the Pentagon but became preoccupied in his first year with the issues of "don't ask, don't tell" and expanding combat roles for women. He commissioned the Bottom-Up Review to back into a force structure justification for the steep budget reduction he imposed on the department, but he left the work largely to the joint chiefs of staff.

Conditions were being set that would engender an Air Force Global Hawk and subsequently bring forth its Navy offspring, Triton. A look back to the geopolitical state of affairs in the early 1990s reminds the reader of the profound changes that U.S. defense policy was then undergoing. The Cold War had been won. A new presidential administration arrived in the United States heading a political party that had been out of power for the preceding twelve years, with a mandate to look inward to long-simmering domestic issues. Reimagining a bold new defense posture that eschewed traditional big-ticket defense procurements in favor of cheaper, less manpower-intensive new technologies was the mood of the moment. To gain a deeper understanding of how, in that environment, conditions were set to advance unmanned aerial vehicles, it is essential to explore the mechanisms used by the incoming administration's Department of Defense—the processes, commissions, and reviews employed to bend the consensus toward a predetermined outcome.

The Bottom-Up Review would reset defense budget priorities, an adjustment that curtailed many ongoing weapons projects and forced outright cancellation of many more. Whereas Aspin was the intellectual figurehead of the Clinton defense establishment, his deputy secretary, Bill Perry, was the shop foreman. Perry, a Stanford-educated PhD mathematician, brought with him a well-formed set of experiences and viewpoints on the management of defense acquisition, gained in part by an industry background as a director of Sylvania's Electronics Defense Labs and subsequently as president of Electromagnetic Systems Labs. His industry work was further enriched by government service as under secretary of defense for research and engineering in the Jimmy Carter administration. His breadth of experience and growing reputation for being fairly apolitical earned him a seat in 1986 on a bipartisan panel chartered by President Reagan that would become known as the president's blue ribbon commission on defense management or, more simply, the Packard commission, named for its chairman David Packard, a technology industry icon and former deputy secretary of defense.

The Packard commission unveiled recommendations for deep structural reform of the defense establishment. This was stimulated in no small part by the lack of interservice coordination and outright operational inefficiency and confusion exhibited in Operation Urgent Fury, the invasion of Grenada three years earlier. Charged by President Reagan to investigate the department's organizational and operational relationships—but with a mandate expanded to address the defense acquisition process when headlines broke about $400 hammers[3]—the members convened in the summer of 1985. They submitted the product of their exertions a year later.

In addition to process changes that put defense budgeting on a different rhythm, their report outlined criteria for drafting weapons requirements that were more actionable; milestone-based authorization for greater in-progress oversight; off-the-shelf procurement of commercially available products; and multiyear procurement of major defense systems. Organizationally, the panelists concurred in strengthening the authority of the chairman of the joint chiefs in his advisory role to the secretary of defense and the president and in elevating the office of the vice chairman as the focal point for integrating often conflicting views among service chiefs and combatant commanders. To corral the many disparate and mostly systemic problems with acquisition, the commissioners also recommended the creation of a new under secretary of defense for acquisition, granting the office authority in acquisition matters over newly created service positions for acquisition executives. The commission's recommendations were codified by Congress by an overwhelming plurality of Democratic and Republican votes that yielded the Goldwater-Nichols Department of Defense Reorganization Act in October 1986. Largely left to future regulatory mandates were commission deliberations on streamlining technology transition.

This was an unsatisfactory omission for Perry, who believed that the commission fell short in not tackling processes for more efficient grafting of systems-level innovation onto defense acquisition. Perry was concerned about a pattern of starving new prototypes of needed seed money due to single-service bias or internecine squabbling when two or more military services were involved. He insisted on wresting day-to-day control of systems-level prototyping from the services and entrusting that to a jointly staffed program management structure within OSD. It was not surprising, then, that he moved quickly in his first year in office as the deputy secretary in 1993 to remake the prototyping process. He brought in key people to flesh out the mechanics and by early the next year had endorsed a new model for rapid, cost-effective demonstration of promising systems-level technologies.

The centerpiece of this new model was the advanced concept technology demonstration, which begat Air Force Global Hawk, which transformed into Navy BAMS, subsequently named Triton. The ACTD had its roots in the unfinished business of the Packard commission. Management of ACTDs was typically to be vested in DARPA and, in the case of those earmarked for identified ISR shortfalls, given to the Defense Airborne Reconnaissance Office (DARO). Its first appointed director, Air Force Gen. Ken Israel, was expected to coordinate closely with DARPA in judging their utility when tested.

Historical problems with systems-level technology transitions failing to meet projections were seen as symptomatic of overly pressurized expectations and onerous contract stipulations, particularly in past UAV programs. DARPA was allowed to deviate from standard Pentagon acquisition processes in several important ways, as exemplified by "use of Section 845 Other Transaction Authority (OTA); use of Integrated Product and Process Development protocol; a small joint program office; user participation through early operational demonstrations; and a single requirement—unit flyaway price, determined as the cost of production and tooling essential for building a single unit or aircraft—with all other performance characteristics stated as goals."[4]

Section 845 language, taken up in 1993 in the armed services conference committee mark-up and passed in the 1994 National Defense Authorization Act, permitted DOD to deviate from standard contracting procedures for technology development, specifically granting DARPA authority to enter into prototype projects using nonstandard contracting. Section 845 benefits included "lower overhead costs by eliminating complex reporting and auditing processes. Additionally, decisions were made faster because the contractors owned the decision processes."[5] The language of Section 845 thoughtfully introduced a "sweetener" for industry to stimulate greater investment in the outcome of prototype development by treating contributions as independent research and development, a portion of which was recoverable by the industry contractor as allowable overhead charges.

Larry Lynn, an insider with a bent toward reforming acquisition practices within the department, probably had the greatest direct influence on designing the contours of the ACTD program. He was appointed deputy under secretary of defense for advanced technology in the newly installed Les Aspin/Bill Perry hierarchy. No stranger to confronting and overcoming the bureaucratic inertia that often stymied the transition of promising technologies, Lynn had served in a previous administration as director of defense systems reporting to the under secretary for research and engineering. He next was a deputy director of DARPA just as the Packard commission was building a head of steam in its 1985 formation and deliberations. Lynn's work in 1993 as co-chair on the Defense Science Board study on global surveillance and his subsequent recommendations as the department's advanced technology leader created the conditions from which Global Hawk would emerge. Those recommendations resulted in the split of the highest capability tier of UAVs into two separate, more manageable paths defined by Tier II+ and Tier III–. The only proviso was that they come under the control of a centralized authority that transcended any single military service's control. Pentagon leaders endorsed Lynn's report and went further by approving DARPA control of all UAV ACTDs: "DARPA was tasked to flesh out a system concept for the Tier II+ while maintaining an aggressive unit flyaway price goal of $10 million. The ACTD process provided for streamlined management and oversight, early participation by the user community, and a tight, disciplined schedule. It also permitted advanced operational prototypes to be sent to DOD's regional warfighting commands for experimental use."[6] Lynn's ACTD precepts seemed to offer an attractive alternative to the standard acquisition whereby now "industry could respond to military needs and requirements by using commercially available technology to build new systems in just years instead of decades . . . allowing new sensors or weapons already in existence to be integrated quickly."[7]

To ensure Tier II+ benefitted from disciplined, experienced oversight, Lynn enlisted close associate John Entzminger to join DARPA as the program manager. Entzminger architected a program plan that could be shoehorned under the flyaway cost cap by calculating cost only on the second planned year of production: "By assessing only the second set (units 11–20) of ten aircraft produced, he could center the $10 million cost goal on a Fiscal Year 1994 dollar based unit flyaway price. . . . Anything not flying, such as spare parts, ground-stations, air crews and support teams, maintenance and logistics, and so on would not be included in the $10 million figure. . . . Here, the seeds were laid for what would become Global Hawk."[8]

By June 1994 DARPA moved ahead on the HAE UAV with release of the request for proposal, the preamble of which made clear to prospective bidders that here was something completely different in the way of defense system solicitations: "This solicitation is a radical departure from the normal Defense Department's request for proposal. It is different because the offerors are being asked to propose their own unique program to a set of objectives. . . . Our objective is to gain insight into your streamlined technical and management approaches which will be used to achieve our Unit Flyaway Price requirements."[9] Industry was invited to consider Tier II+ vehicle top-level performance requirements that included an ability to "loiter over the target area for 24 hours at an altitude of about 65,000 feet; a mission equipment package that should consist of a synthetic aperture radar and/or an electro-optic/infrared sensor, a data recorder subsystem, a threat warning receiver subsystem, and an airborne data link subsystem that would transmit data to the ground station that, in turn, would synthesize and display sensor data."[10]

None of these requirements were deemed strictly essential; the only bottom-line mandate that had to be complied with was the unit flyaway price of $10 million. According to the DARPA solicitation, the ACTD for the Tier II+ HAE UAV would progress through four stages. Three selectees would enter a six-month initial design phase that would result in two

High-Altitude, Long-Endurance UAV Performance Objectives

Characteristics	Conventional HAE UAV
On-station loiter	24 hours
Operating radius	2,000–3,000 nautical miles
Loiter altitude	60,000–65,000 feet above mean sea level
True air speed	300–375 knots
Takeoff weight	15,000–27,000 pounds
Survivability	Threat warning, electronic countermeasures, decoys
Sensor payload	Synthetic aperture radar, ground moving target indicator, electro-optic/infrared sensing
Command and Control	UHF FLEETSATCOM
Ground Control	Max use of government/contractor off-the-shelf

Source: DARPA HAE UAV ACTD Management Plan, Version 1.0, December 1994 (draft).

finalists being chosen to enter phase two, a twenty-seven-month period to complete detailed design and produce two prototype air vehicles, a sensor suite, and a ground station suitable for flight testing. A single contractor would be then selected to enter the third phase of thirty-six months, at the end of which eight preproduction air vehicles and two ground segments would be delivered and an operational demonstration completed. With successful completion of the third phase, the contractor, in phase four, would be offered a lot I contract for ten fully integrated air vehicles and two ground segments for the recurring flyaway price of $10 million per vehicle. Encouraged by the breadth of innovation reflected in the proposals by the fourteen respondents, DARPA elected to enter phase one with five designs, those entrants each receiving contracts for $4 million.[11] Among them was Ryan Aeronautical in San Diego, a unit of the expansive Teledyne conglomerate, which submitted a concept that closely resembled what would become Global Hawk.

Few of the 50,000 passengers that pass through San Diego's international airport daily probably know it by its historic name, Lindbergh Field. Fewer still are likely aware that one of the earliest tenants that sat on the outlying marshland by the airport perimeter was tiny Ryan Airlines, manufacturer of Charles Lindbergh's *Spirit of St. Louis*. The Ryan Airlines of that day bore only a distant relationship to the Ryan Aeronautical Company of 1994, mainly as one of the several Ryan companies founded by energetic entrepreneur and businessman T. Claude Ryan in 1934 during the raucous years of a young aviation industry. Sixty years later, Ryan Aeronautical had narrowed its product line to specialize in unmanned reconnaissance drones. To be sure, these drones were still very capable flying vehicles but were generally of smaller size than manned aircraft and, in the case of target drones, expendable. Along the way, Teledyne had absorbed the boutique aerospace manufacturer in 1969, but the name Ryan was retained because it still struck some resonant chord among many working in the aviation industry.

Ryan's president, Bob Mitchell, was in the hunt for opportunities that could be the springboard for renewed growth in a highly competitive market space for unmanned systems. Ryan's fortunes had crested in the Vietnam War era with the cancellation of the company's multiservice

BQM-145A, a medium-range reconnaissance aerial drone, and the fresh loss of a medium-altitude, long-endurance UAV competition to cross-town rival General Atomics, added some urgency to his search. Elevated from the ranks of Teledyne Brown Engineering to run Ryan in the late 1980s, Mitchell began his early working career as a flying officer in the Royal Air Force and later attended the U.S. Air Force Institute of Technology, earning a master's degree with distinction in astronautical engineering. Mitchell possessed a prodigious faculty for not only identifying and organizing exceptional engineering talent, but also divining his customer's desired end state—usually an exercise in prioritizing a value proposition that placed an offering in the best light for competitive selection. As would be shown again in the BAMS contest, he also had nurtured an almost instinctive command of price-to-win strategy and had shown repeatedly how to adapt it to different environments, products, and customers. Mitchell knew how to win, and he was a fierce competitor. He knew the numbers of the Navy budget, what the potential usable budget was for internal cost and pricing of a bid, and where to establish the price-to-win.

One means Mitchell employed to aid him in this latter regard was to bounce his design approach and strategy off a panel of esteemed retired senior military officers for their reactions and recommendations. He worked known contacts in the larger customer community directly and through Ryan's field offices, particularly in Washington, DC. A strategy for mobilizing influence included a congressional contact plan that was plotted by his Washington office employees and consultants. Internally to his team in San Diego, he appeared very much the traditionalist leader that would emerge as an important enabling presence for BAMS in the Navy. At this moment, and again years later in the BAMS competition, Mitchell exhibited an empowering leadership style, granting license to those individuals exhibiting the innovative thinking he wanted to create a defined value proposition. He repositioned people and resources to create a Tier II+ division and staffed it in quasi-secret fashion. Chief engineer Hermann Altmann, structures lead Alfredo Ramirez, and a small hand-picked team of innovators were bound by few constraints on their creativity save obeisance to the $10 million unit flyaway price. The team was given free rein to work issues openly across disciplines within the confines of the Tier II+ cell—mimicking the intent of an integrated product and process development protocol for integrating acquisition activities through multidisciplinary teams. Mitchell had the team map complete system and subsystem interfaces and perform functional checks in the laboratory. With a nod of approval from Air Force technicians at Beale Air Force Base, linkages were established to the Ryan lab to replicate an intelligence system hookup, with the Air Force providing live feedback and recommendations.[12]

The Tier II+ design trade space drew on the full portfolio of Ryan Aeronautical's existing and prior unmanned system experience. So, for example, the wing and landing gear were borrowed from its Compass Cope design, ground control console design from the Lightning Bug, and engines from an off-the-shelf commercial design flying on the Citation X business jet produced by Cessna.[13] The payload packaging relied on Hughes Aircraft Company—eventually acquired by Raytheon—for the electro-optical and infrared camera and a derivative of a widely marketed radar for mounting on the ventral surface. For satellite communication with the aircraft, a four-foot dish could be mounted in a frontal dorsal lobe that operated in the Ku frequency band, generally available worldwide at the time. The engine inlet required some care in design to eliminate any obstruction of antenna-to-satellite signal reception.[14]

U.S. government and military officials have chosen a flying wing and four more conventional wing-and-tail aircraft designs as finalists in the competition for the $230 million Tier 2+ unmanned aerial vehicle program. The winners were: Raytheon, which teamed with Lockheed's advanced development company to offer a flying wing. This comes as little surprise since Lockheed designed flying wings for the stealthy Tier 3- and much larger, but now canceled, Tier 3 UAV programs. Other team members include TASC and Allied Signal Aerospace; Loral Western Development Labs, teamed with Frontier Systems, Loral Defense, and Loral Fairchild; Northrop Grumman with Westinghouse Electric, AeroVironment, which built the Raptor Pathfinder solar-powered UAV, Burt Rutan's Scaled Composites and PAR Government Systems; Orbital Sciences teamed with Scaled Composites and Westinghouse; and Teledyne Ryan Aeronautical, which formed a team with two E-Systems divisions—Melpar and Garland—GDE and Adroit Systems.

—DAVID FULGHUM, "Flying Wing, Four Others Selected for Tier 2+ UAV,"
Aviation Week & Space Technology, October 10, 1994

As they digested the news that Ryan was an awardee for the initial design, Mitchell and his small team of free-wheeling configurators began to grasp that they might have a real shot at outlasting the competitive field of larger, better resourced aerospace companies to bring home the ACTD for the third phase of user evaluation. A renewed sense of purpose and urgency overtook them, inspired by Mitchell's upbeat persona and nonstop encouragement. When DARPA took a budget cut that reduced monies available for the second phase, the government evaluators decided they had seen enough of the industry designs to make an informed down-select to just one contractor team (rather than two as originally planned by John Entzminger) to enter the second phase for prototype build. Ryan Aeronautical, and its UAV, was their selectee in May 1995. Under Mitchell's vision and guidance, Ryan had made a compelling case that it could best satisfy the performance demands and had the cost control mechanisms in place to make the unit flyaway price.

Early in the second phase, potential names were submitted in response to a naming contest hosted by DARPA, with Gen. John Ralston, then Air Combat Commander, selecting the winning name—Global Hawk. An icon in aviation lore was born. But a long and tortuous road remained to bring Global Hawk to full operational maturity. Only in his few quiet moments could Mitchell allow himself to even contemplate the future business potential and growth in market UAV share; all energy and every hour had to be focused on performing to DARPA's expectations.

Air vehicle design-and-build proceeded relatively smoothly. Among the notable innovations brought to the design was the routing of super-cooled fuel at altitude around the heat-producing electronic components to remove heat—the heated fuel was prevented from seizing in the fuel lines by cold-soaking for extended periods at 65,000 feet. But the Ryan design team was vexed by problems with the mission computer software, causing more than a yearlong slip in the schedule. Despite experiencing the minor glitches that emerged during most initial flights of a new-design aircraft, Global Hawk completed its first flight successfully on February 28, 1998. Air vehicle number one climbed to 32,000 feet and remaining aloft for 56 minutes. Air vehicle number two was not to experience the same good fortune, flailing to an ignominious impact and destruction on the California desert floor a year later. According to the Air Force accident investigation, "The mishap occurred when Global Hawk inadvertently received a signal commanding flight termination from a test range at Nellis Air Force Base, Nevada, which was outside the frequency coordination zone in which the UAV's mission was being flown. This caused the Global Hawk to go into a termination maneuver involving a pre-programmed, rolling, vertical descent from an altitude of 41,000 feet."[15]

The sterile and flat technospeak of the report masked some genuine discouragement and concern at Ryan. The test schedule had already been reprofiled, which brought heightened scrutiny to the ongoing evaluation of the one remaining aircraft. The increased scrutiny surely prompted speculation among Ryan employees about more bad news to come. Naysayers would soon emerge to question the wisdom of this unmanned aircraft experiment. Furthermore, the use of unconventional new pilot acquisition provisions allowed as other transaction authority were openly questioned. Scrutiny centered on the extraordinary latitude the OTA gave Ryan by transferring significant design and test responsibility to the contractor. The question of whether Ryan was up to the job no doubt plagued Air Force acquisition officials—and probably many at Ryan as well. The official incident report went on to suggest that disregard of established procedures and lack of methodical coordination between the Air Force and Ryan personnel contributed to the crash, but that it was "not clear if the contractor's designation as lead for the test program execution played a role in the loss of Air Vehicle #2."[16]

What was originally planned as a twelve-month flight test period within the twenty-seven-month-long second phase slipped by four more months, due partly to the loss of air vehicle number two. That loss precluded characterization of the electro-optic/infrared camera payload performance in the time allotted. Nonetheless, satisfactory airworthiness was confirmed. Air vehicle, sensor, and software issues aside, a more deep-rooted issue became exposed when management of the program transitioned from DARPA to the Air Force. Because of the schedule delays, the transition, originally planned to occur at the end of the second phase, was delayed until October 1998. Transition then occurred prematurely, prior to the second phase completing, with only five of sixteen test flights accomplished. Flight test discrepancy reports were piling up, and the Air Force received management responsibility for an acquisition program the likes of which could not be recognized outside of DARPA.

In truth, a feature of the ACTD built into the program structure by Entzminger was an inordinate amount of control acceded to the contractor from what would normally be expected in a routine defense acquisition. Under Section 845, the conventional rules had been waived in order to streamline the ACTD acquisition and shield it to the greatest extent from the normal bureaucratic burdens. This streamlining earned its own designation in the argot of new, experimental defense acquisition oversight: total system performance responsibility. As a result of its inclusion in the ACTD formula, the managing government program office could remain lean and, hopefully, agile, as long as Ryan, in this case, was meeting the minimum performance thresholds, holding reasonably well to schedule, and complying with the $10 million flyaway price. This gave Ryan near-absolute program authority with very limited direction from the government program office, absent any formal reporting or tracking.[17]

Upon transition of the ACTD, no forewarning or advance coordination had been accomplished with the Air Force to prepare the service for taking on a highly unorthodox program and somehow making it conform to the Air Force acquisition system. This was the case of the unstoppable momentum of the ACTD meeting the immovable object of the Air Force bureaucracy. Compounding this shock to the Air Force system was the lack of clear ownership; the aircraft and its supporting systems required a sponsoring command that desired the capability and would stand up to defend its resource needs—in other words, the user. Initially, this fell to the Air Combat Command headquartered in Langley, Virginia, where it would have been difficult to discern anything close to full-throated advocacy for the program. At Air Combat Command, "the staff did not feel enamored with the program."[18] Officers on the headquarters staff "knew the senior Pentagon leadership intended to show that this Global Hawk thing worked and then push it on the Air Force. . . . Since it wasn't developed within the Air Force requirements process, Air Combat Command staff officers were very resistant to the concept."[19]

Here was an irony playing out that might have been anticipated at the formation of the high-altitude, long-endurance UAV ACTD: steps taken to streamline the acquisition and protect it from the encumbrances of the Pentagon bureaucracy actually intensified its troubles when it came to reintegrating into the formal acquisition process by introducing an unfamiliar construct into a highly regulated and disciplined rule set without the necessary step of building awareness and support. The third phase of the ACTD began almost immediately upon completion of the second phase in June 1999. Air vehicles three, four, and five were delivered to populate the initial inventory, intended for demonstrating Global Hawk utility in a set of exercises and evaluations. In the Air Force, some associated with the program doubted that tiny Ryan would be able to keep pace with a potential downstream transition to full-rate production, but for the moment, attention focused on exercise planning with the available aircraft and ground control systems.

As 1999 closed, Ryan Aeronautical was ensnared in the surge of industry consolidation, absorbed by one of the last large defense giants. Six years before, in that seminal year of 1993, the industry consolidation rush initially set in motion by Les Aspin was continued by Bill Perry when he invited fifteen defense industry chief executive officers to the Last Supper in the Pentagon. The logic laid out for the executives that evening sprang from an accepted truth that, with the end of the Cold War and the U.S. standing in the world unchallenged, defense budgets could be expected to decline by at least 40 percent. They would not experience a return to anything like pre-1980s spending levels. The reasoning served up to the executives was that without a corresponding retrenchment by these companies in plant and equipment, lower defense spending would equate to higher unit costs that would be borne by industry. A profit squeeze would likely force some—possibly many—out of the industry, and with them a loss of experienced defense workers and valued plant and equipment that were considered vital to retain.

That night, the discussion was purposely scripted to force a more thoughtful downsizing of the industry and "encourage the companies, through normal capital market mechanisms, to make rational business decisions that would result in fewer assets devoted to defense."[20] The

Department of Defense sought to show the way through an aggressive downsizing of its own by means of the base realignment and closure process. Bases, depots, and laboratories around the country were shuttered and their real estate holdings surrendered to local municipalities once any toxic waste remediation was provided for.

The Last Supper spurred a wave of consolidation that would run strong for five years: In that first year after the meeting, Lockheed purchased General Dynamics' aircraft division; Martin Marietta overtook General Electric's aerospace unit; Loral picked up IBM's federal systems, followed by Martin Marietta's purchase of General Dynamics' space launch vehicles. In 1995 Martin Marietta merged with Lockheed to form Lockheed Martin, and General Dynamics acquired Bath Iron Works. In 1996 Lockheed Martin completed its acquisition of Loral, and Boeing acquired Rockwell. The following year, Boeing acquired McDonnell Douglas, and near the end of this merger spree, Raytheon picked up Texas Instruments and Hughes.[21]

In the midst of this merger wave, Northrop's chief executive, Kent Kresa, set about remaking the company he had headed since 1990. Since arriving at Northrop fifteen years earlier, Kresa had built on a recognized competence in both management and technical positions to quickly ascend to president and chief operating officer by 1987. He emerged as the odds-on favorite to replace longtime Northrop chairman Tom Jones upon his retirement in 1990. By 1993 Kresa had helmed the company through three years of declining prospects, albeit ones not entirely of his making. Production of Northrop's most recognizable marquee product from the 1960s through the 1990s—the F-5 Tiger fighter aircraft and its cousin, the T-38 Talon trainer—had already ceased; the company's prototype YF-17 had come up short in the Air Force's lightweight fighter program in head-to-head competition with the F-16, and when the YF-17 prototype was resurrected for the naval fighter attack experimental program, Northrop's position as prime was transferred via the legally binding teaming agreement to McDonnell Douglas—although Northrop retained some 40 percent of program content as a supplier of subassemblies to the F/A-18.

One year into Kresa's tenure, Northrop's YF-23 prototype stealthy fighter aircraft entry in the Air Force's advanced tactical fighter program lost out to Lockheed's YF-22, and the company's stealthy tri-service standoff attack missile development was spiraling into cost and schedule overruns that would eventually lead to its cancellation. Two years into his tenure, Northrop's B-2 program, already declining in numbers from an initial inventory projection of 132 aircraft, fell further to just 20 production aircraft. Compounding Kresa's problem with a steeply declining production base were the lingering effects of a criminal indictment of Northrop by the Justice Department in 1989 for falsifying test certifications for missile guidance parts. Under his leadership, restitution was made out of court, avoiding a further prolonged battle with the Justice Department. But even with a recognized and historically significant aerospace marquis and a ranking among a shrinking handful of legacy defense companies left in the U.S. market, Northrop was bleeding cash, and breakup or acquisition looked increasingly likely.

Kresa recognized that a new strategic direction was needed to reverse the slide. An astute observer of military trends, he discerned a fundamentally new doctrine emerging in writings and policy documents from the defense establishment that were gaining new adherents in the senior officer ranks under the heading of revolution in military affairs (RMA). In short, this was the waging of war by leveraging superior information and precision effects to counter asymmetric threats emerging on the world stage. Successful application of these revolutionary effects depended on a set of undergirding technologies in ISR, in command, control, and communications, and in sensors, networks, and unmanned systems. Here was the new strategic direction in which Northrop might pivot while simultaneously extricating itself from near-total dependence upon a declining number of make-or-break advanced fighter and bomber industry competitions. To Kresa, Les Aspin's Last Supper appeared to be exactly the sanctioned endorsement—even the encouragement and top cover—needed to set a new vision and permit a rebuilding of Northrop through aggressive mergers and acquisitions.

Kresa tapped advisors and consultants inside and outside Northrop to assess the implications of the Last Supper and to recommend next steps for him and the company in light of those implications. A convincing analysis from his outside consultants confirmed Kresa's thinking that Aspin's Last

Supper presented to Northrop three different courses of action: "eat"—either obtain diversification by acquiring companies that were largely operating in civilian commercial markets to ameliorate downturns in defense, or acquire another large defense company (or companies) to gain product diversification and market share; "be eaten"—surrender Northrop's balance sheet and its independent future to a larger buyer but gain shareholder value in a good sale; or "close up shop"—liquidate nonperforming defense units in piecemeal fashion while looking for entry points to grow into commercial markets. His consultants advised that the most attractive option appeared to be "eating" another large defense company (or companies) to help Northrop achieve penetration of the emerging RMA market. Kresa agreed with his consultants' recommendations and understood that they could yield an outcome giving Northrop the scale to effectively compete in a shrinking field of industry competitors.

Another attendee at the Last Supper was Renso Caporali, who had taken over Grumman in the same year that Kresa was elevated to the top job at Northrop. Under Caporali, Grumman, like Northrop, was experiencing a shrinking business base in aircraft production. At the time, defense stock analyst William Deatherage observed of Grumman: "All of their aircraft programs are in trouble."[22] In the year before Caporali's elevation to chief executive, Secretary of Defense Dick Cheney had attempted to zero the funding for further production of the company's flagship program, the F-14 Tomcat—then the Navy's premier fleet defense fighter aircraft—despite a vigorous lobbying effort by the company.

A year later, Cheney, annoyed with Grumman's aggressive lobbying, cancelled further upgrades to its premier long-range penetrating attack aircraft, the A-6 Intruder. In its place, the Navy initiated the AX (attack experimental) program in 1991. Grumman participated as a team member with Lockheed and Boeing, competing against four other industry teams. When the separate but parallel naval advanced tactical fighter was terminated, some of its requirements were absorbed by the AX program, and it quickly evolved into the A/F-X (attack/fighter experimental aircraft). Grumman and its AX teammates adjusted and bore down on new A/F-X requirements. Then, in the first year of Secretary Aspin's tenure, his Bottom-Up Review and the results of a DSB study converged in 1993 to prompt DOD to cancel the A/F-X for projected ballooning cost and lack of an executable program plan on the part of the Navy. With the demise of the A/F-X program, DOD's tactical aircraft roadmap was redrawn around Lockheed's F-22 for the Air Force and F/A-18E/F for the Navy, with future development focused on a nascent joint advanced strike technology program. It was not a new aircraft design effort in the traditional sense, but a collection of technology contracts that folded into the joint strike fighter aircraft development. Unknowable in 1993 was that the joint strike fighter would gestate for two more decades before being declared operationally ready by its first user. Events in 1993 would unfold to profoundly affect Grumman's autonomy; the company's future as a prime aircraft manufacturer effectively ended that year.

From the West Coast, Kent Kresa and his mergers and acquisitions staff at Northrop observed something else about Grumman. Even as the future looked dim for a company renowned since the 1930s as a mainstay producer of naval aircraft, Grumman also held preeminent positions in key niche markets for ISR aircraft. With its EA-6B Prowler electronic warfare aircraft for the Navy and EF-111 Raven jamming aircraft for the Air Force, Grumman possessed a near-monopoly on the black art of airborne electronic attack. Even though initial production of both had run its course, a thorough understanding of the arcana of this mission area unique to Grumman gave the company a distinct competitive advantage in any follow-on contest for upgraded aircraft or recapitalized inventory. Grumman's large four-engine modified Boeing 707 passenger jet, the E-8 joint surveillance and target attack radar system ground surveillance and battle management aircraft, had recently acquitted itself flying in its preproduction development phase configuration in Operation Desert Storm, and prospects for sustained production appeared strong.

Airborne early warning for Navy aircraft carriers was another monopoly position for Grumman with its carrier-based E-2 Hawkeye airborne early warning aircraft. In 1993 upgrades to the E-2 were continuing, with the potential for many more years of major production line modifications. In a defense environment increasingly swayed by the so-called revolution in military affairs, Grumman's stable of ISR thoroughbreds fit in an unusual, protected category of low-density/high-demand assets. Modification and reprocurement budgets were much more likely to withstand downward pressure for these highly prized, and more rare, special mission aircraft. Yet the strength of these programs would not yield revenue adequate to sustain Grumman as a prime manufacturer. By their nature, they belonged to a class of warfighting machines that was limited in number and would never enjoy long production runs. Grumman, under Caporali, recognized its inevitable decline and began private internal analysis of potential merger candidates. Secret discussions with Northrop about a possible merger of near-equals that might yield critical mass were begun and progressed favorably for almost a year.

Nearing a final board decision in Bethpage to proceed with the merger, Grumman's investment bankers threw a wrench in the proceedings by suggesting the company also investigate a possible combination with Martin Marietta. Kresa and Northrop were not alone in perceiving value in Grumman. Martin Marietta head Norm Augustine and his corporate board authorized an initial bid of $55 per share for Grumman.[23] What had started a year before as friendly merger discussions with Northrop now turned into a much more hostile acquisition bidding war between Northrop and Martin Marietta that some believe was provoked by investment banker Goldman Sachs to drive up their fee. Martin Marietta hoped to score a quick buy by making its offer in cash rather than stock. Grumman's board responded favorably to the offer, agreeing to be acquired. But Grumman's directors, upon hearing Northrop's final bid of $62 per share, reversed themselves and gave in to the inevitable.

Grumman Corp. accepted Northrop Corp.'s $2.17 billion merger offer yesterday, capping a month-long takeover battle and ending Martin Marietta Corp.'s attempts to buy the Bethpage, N.Y.-based defense contractor. Northrop's chairman, Kent Kresa, said yesterday that he anticipates a smooth transition and reaffirmed his earlier pledge of having the combined companies' name changed to Northrop Grumman Corporation. Mr. Kresa said a merger of Northrop and Grumman will create a world-class military aircraft design capability in both tactical and surveillance aircraft.

—TED SHELSBY, "Grumman Says 'Yes' to Northrop," *The Baltimore Sun*, April 5, 1994

The new Northrop Grumman seemed poised to elbow its way to the front rank in the burgeoning emerging technologies market of the post–Cold War era. More acquisitions were to come, but none defined the reborn corporate giant like the joining of two such highly regarded defense legacies as Northrop and Grumman. The name itself implied a synergistic combination of two powerhouses in the history of U.S. aerospace giants. But this was not a marriage of equals as the name suggests—it was very much a hostile takeover that had turned acrimonious in the bidding war for Grumman. In the months following the acquisition, "Northrop executives only paid lip service to the idea that integrating two cultures would be the biggest challenge."[24] Some years later, a Northrop Grumman employee with significant exposure to the company's merger and acquisition history recalled the experience of the Grumman takeover:

> Never underestimate the culture of the company. When I come across people from Grumman, they will still complain and moan about how Northrop did them wrong—it's been 17 years! Grumman and Northrop were both old aerospace companies that were very similar companies, but one was on Long Island and one was in California. If you worked on Long Island, you were employed by Grumman. It owned Long Island—they gave away turkeys to employees for Christmas. They were trucked onto Long Island and everyone would line up—retirees, employees, everybody would line up and get their turkey. One of the first questions asked during our first meeting with them was, "Will we still get our turkeys?" These cultural icons are really important to note and celebrate in some way, even though you aren't going to keep them. You don't just come in there and trash the business.[25]

But from the view of Grumman's rank-and-file employees, that is what happened. There was a strongly held belief on Long Island that a cherished family atmosphere among employees and their supervisors had been cavalierly abandoned and that a venerated tradition of customer intimacy nurtured over sixty years (especially with naval aviation) had been trashed by the new owners in a most abrupt and disdainful way.

Northrop anointed a senior officer from its own executive ranks in California, Ollie Boileau, to relocate to Bethpage and assume leadership over the Grumman unit with directions to shake up the organization, reform staid processes, and generally bring Grumman's recalcitrant culture to heel. Boileau's leadership style was characterized, in whispers around Grumman hallways and watercoolers, as "take no prisoners and shoot the wounded." His reputation transferred generally to Northrop employees in the minds of the Grumman work force. Northrop was soon regarded as "those imperious West Coast technocrats." Resentment would smolder for years to come; it would flare even ten years later to have a decidedly negative impact on the BAMS program.

In the last months of Grumman autonomy before Northrop's takeover, the advanced design staff felt extremely confident that they were hitting all the right notes with their government counterparts at the Naval Air Systems Command on A/F-X preliminary design work for the Navy's next experimental fighter attack aircraft. Even with the Long Island workforce in a steepening decline, the company had wisely retained a core of aerodynamicists, loads engineers, carrier suitability designers, operations and mission analysts, and, importantly, the models, routines, and labs that supported their work and that had established unquestioned credibility with Navy evaluators. That credibility went deeper than self-serving company advertisements or executive PowerPoint "who we are" presentations. It was nurtured and grown over the previous sixty years by personal relationships between professional counterparts and by Grumman meeting its commitments, not overpromising, and almost always doing the right thing for the customer. The result was a successful run of Navy tactical aircraft production lines interrupted only by the periodic but recurring intrusion of another legacy Navy aircraft producer, LTV, better known by its pre-1963 moniker, Vought. Save more infrequent appearances by other occasional Navy producers such as Curtiss, Douglas/McDonnell Douglas, and Boeing, Grumman and Vought seemed to swap leads as the dominant Navy producer from the beginning. Vought, older by date of founding, led in the 1920s with production of the VE-7, the first aircraft to take off from the Navy's aircraft carrier, followed by the UO-1 and O2U biplane fighters.

Grumman overtook Vought in the 1930s with more improved biplanes and then the Navy's first single-wing fighter, the F4F Wildcat. Vought began a return to the top rank in the late 1930s with the SB2U Vindicator dive-bomber. Grumman's F6F Hellcat debuted in the early years of World War II to rule the skies over the Pacific from 1943 through 1945. By the end of the war, Vought with its F4U Corsair challenged the Hellcat's reign. The Corsair largely supplanted the Hellcat in Navy air groups until the early jet age, when the Corsair gave way to Grumman's F9F Panther and swept-wing Cougar turbojet-powered fighters. Within just a few years, Vought overtook Grumman's jet cats with its F8U Crusader. Excepting the McDonnell Douglas F-4 Phantom in the 1960s through early 1970s, Grumman again surged ahead with the F-14 Tomcat, and the two companies vied for dominance in the attack class, Vought with its A-7 Corsair II and Grumman with the A-6 Intruder.

From the inside, much about Vought's inner workings closely resembled those of Grumman. Both had arrived in the 1990s long committed to vigorous campaigns to cultivate relationships with NAVAIR counterparts—most especially George Spangenberg, head NAVAIR evaluator, but also others in the functional divisions, such as Mike Dubberly and Don Palokovics in structures and configuration design and Tom Lawrence in aero and flight clearance. Both companies had retained strong front-end design staffs—at least through the end of the 1980s—that possessed, uniquely among U.S. aircraft manufacturers except for McDonnell Douglas, the carrier suitability designers, the special analysis routines on which the Navy relied, and an underlying strength in Navy mission analysis and operational employment schemes. At Grumman, especially, the industry models routinely used for playing out engagements between attackers and defenders to characterize effectiveness of air defense artillery, surface-to-air missiles, and low-altitude radar systems against carrier-borne aircraft were staples.

More relevant to characterizing maritime surveillance, the Grumman simulation staff was highly adept at developing mission-level outcomes using the somewhat cumbersome and laborious naval simulation system model, generated by the Space and Naval Warfare Systems Command, and Grumman's own multi-aircraft penetration and engagement simulation model, derived from earlier F-14 chainsaw tactics development and highly regarded by evaluators at the Naval Air Systems Command. These latter two routines would be particularly useful in dramatizing the effectiveness of a BAMS-like surrogate UAV in the initial BAMS concept exploration study.

In a bit of irony foretold, Grumman and Vought teamed with prime Northrop in the mid-1980s to design the advanced tactical aircraft, the Navy's A-6 replacement for penetrating long-range strike. Grumman and Vought expertise with Navy design characteristics and mission understanding would prove essential for reducing Northrop's concept of a large, survivable attack aircraft for operation from an aircraft carrier to a more manageable size. The team produced a viable concept that seemed to meet the Navy criteria in all areas except contract type. The Navy insisted on a fixed-cost incentive-type contract as a condition for entering full-scale development, and the Northrop/Grumman/Vought team elected to back out. The winners of that competition, General Dynamics and McDonnell Douglas, would later fail to deliver satisfactory contract and design performance with their A-12 concept. Vought and Grumman both reengaged—but on separate teams—in the early 1990s to work on the Navy's next attempt to design an A-6 long-range penetrating attack aircraft replacement, but by then Vought was in extremis. The parent company, LTV Corporation, filed for bankruptcy protection in 1986, which cast doubt on Vought's viability going forward. Critical mass as a financial entity had diminished to the extent that the company was hard pressed to put together the resources that would make it an attractive teaming partner. The A-7, Vought's last new production airplane, had completed its production run a decade before and was on a sunset plan for retirement in both Navy and Air Guard inventories.

Vought had settled into a nonprime role, building large subassemblies for the C-17 and the B-2, even adopting the slogan "preferred partner" in its marketing literature and presentations. The company found it increasingly difficult to maintain complex system integration labs, wind tunnels, large autoclaves, and state-of-the-art composite layup machines and still remain cost-competitive in its new nonprime position as an aerostructures producer in a second-tier market. The company dropped to twentieth-largest DOD contractor from the position in the top ten that it had held throughout the 1950s, 1960s, and into the 1970s.[26] Much-reduced annual sales of $1 billion in 1991 shrank the workforce to fewer than 7,500 employees, and the number was on a downward spiral. Sales dropped again the next year with no new production programs in sight. Vought's five-year backlog of $5 billion on large subassemblies, however, made it an attractive target for takeover.

Not immediately obvious to industry observers, except perhaps those with a discerning eye such as Northrop's chief executive Kent Kresa, was an aspect of Vought's makeup that was retained from headier days. Even in its downward trajectory, the company invested in a core group of configurators and mission analysts that had experience from years before in developing the antisubmarine warfare S-3 carrier-based aircraft as a partner with Lockheed. Again in the late 1980s, Vought's designers fronted for Lockheed with Navy evaluators to innovate an E-2 airborne early warning aircraft replacement based on the S-3 airframe. Vought's innovators were held in the same high regard by NAVAIR as the Grumman designers, at least entering the 1990s.

Partially buoyed by this realization, Northrop considered its next move. In a corporate acquisition that actually predated Les Aspin's Last Supper, Northrop partnered with Carlisle, an investment group specializing in buyouts and the turnaround of distressed corporations, to acquire the Vought Aircraft Products unit of LTV Corporation for $230 million. Northrop, a 49 percent owner, had the option to buy out Carlisle's 51 percent ownership stake inside of three years, but Northrop moved more quickly to consolidate the Vought buy the very next year for $130 million. Compared to the Grumman takeover in 1994, the buyout of what remained of Vought barely registered in aviation circles.[27]

As 1994 ended the new and growing Northrop Grumman portfolio now included two historic, legacy Navy aircraft design businesses, notable for their orientation toward and knowledge of Navy operations, missions, and tactics as well as their intimate familiarity with the Navy's procurement arm and the key evaluators and decisionmakers involved in assessment of aircraft performance and naval warfare utility. None of this would seem especially remarkable or relevant to the development of

BAMS except for the effect that it had in drawing together small focused groups of former Grumman and Vought people, looking for ways to apply experience to strengthen the new enterprise and undoubtedly seeking to establish their own bona fides and relevance during the fast-moving industry consolidation then taking place. Preeminently, this became a feverish exercise in finding a way to insulate oneself from the employee layoffs that picked up steam as units reorganized under the new Northrop Grumman corporate banner.

Despite the abrupt shock of organizational changes wrought by Ollie Boileau, a most fortunate outcome for the former Grumman enterprise was the appointment by the new owners of Paul Bavitz from Bethpage to head Northrop Grumman's advanced technology development center (ATDC—not to be confused with DARPA's advanced capability technology demonstration, or ACTD), the crucible for new and innovative designs. Both Northrop and Grumman operated similar front-end advanced design and innovation units prior to the consolidation, but where Northrop's unit had lost its strategic direction by straying into non-defense products such as the commercially oriented advanced technology bus for urban commuting and the advanced life-sustaining stretcher for battlefield emergency medical response, Grumman's orientation—epitomized by the branch of ATDC at Bethpage—remained resolutely focused on military aircraft products.

Bavitz had been a standout engineering leader whose portfolio included taking the Gulfstream from propeller to jet in one of the first adaptations of super-critical wing design technology to business class aircraft; guiding the Super Tomcat 21 initiative in the face of determined DOD resistance until it was politically buried; and leading Grumman's AX program as prime with subordinate team members Lockheed and Boeing until the competition drifted into A/F-X and was ultimately axed by the Navy due to requirements creep. Bavitz insisted on shoring up a critical mass of ATDC capability in Bethpage to preserve the talent, labs, and experience that he knew would otherwise be allowed to atrophy in deference to Northrop's own ATDC branch unit in El Segundo, California. The resulting organization identity was more like a federated network of innovation centers that held together the remnants of design expertise at both Northrop and Grumman. Thanks in large part to Bavitz's foresight in holding together an innovation and design center in Bethpage, the principal players that would conceive and nurture the BAMS concept from origination to realization were brought into the same orbit.

In 1994, and through 1999, it was not immediately clear just who those principals would be, nor how and when their particular talents and personalities would coalesce to create the impetus for BAMS. For most of that time, Dr. Maris Lapins was building a portfolio of directly applicable experience as the ATDC director of prototyping and simulation, working with advanced technologies for virtual prototyping of new products, distributed simulation, and networking. For the newly established ATDC, he was the go-to authority for virtual and physical prototyping of hardware, systems, and software, an outgrowth of his primary background in aeronautical engineering with a specialty in flight mechanics.

Prior to his arrival in the position, Lapins was an exacting overseer inside the Navy AX program competition (before it morphed into the Navy A/F-X) as director for concept exploration and definition under Paul Bavitz, deftly managing a combined design team of Grumman, Lockheed, and Boeing configurators by a disciplined management style, imposing action item accountability at daily standup meetings and nudging progress through equally frequent functional reviews and issue resolution team meetings. As the BAMS inception drew closer, Lapins had been elevated to director of advanced systems development at the renamed Bethpage ATDC, guiding operations analysis and simulation and advanced design engineering. An experience that rounded out what could best be described as his whole systems approach to concept design was completion of a two-year rotation through Grumman's outstanding performer development program, a selective training regimen for high-potential mid-level managers that provided him hands-on exposure to manufacturing, business operations, and corporate strategy. His opinions normally conveyed a mastery of the engineering disciplines that were a necessary part of creating the whole—a thoughtful systems-level approach to design that conformed to cost and schedule constraints.

Howie Frauenberger had compiled a solid resume of systems engineering and program management experience across a variety of nascent concept design initiatives that seemed to all end in cancellation or loss of the competition. Through each of those experiences, Frauenberger gained a bit more insight and familiarity with the design process, the lessons learned forming in him a more individualistic—even nonconformist—approach toward preliminary design problems. With the dislocation unloosed by the Northrop acquisition, he landed in the Bethpage branch of ATDC, the on-site advanced design center, and immediately took on a systems engineering role on the Northrop Grumman–McDonnell Douglas team's aircraft demonstrator for the joint advanced strike technologies program, precursor to the joint strike fighter. When his team did not progress beyond the government down-select, he found himself redirected to the DARPA/Naval Sea Systems Command arsenal ship competition, where he guided development of operational concepts and very austere, minimum manning schemes. Frauenberger was not accustomed to working ship design programs, but it immediately became apparent that someone with his systems engineering bent would be a natural fit on the team. As the arsenal ship—a Navy concept for a floating magazine of cruise missiles—was about to enter the fabrication phase of a demonstrator ship in 1998, the program was cancelled for lack of funding priority in the Naval Sea Systems Command budget, and he moved on to his next assignment.

Frauenberger had already demonstrated a persistence born of fierce confidence in the rightness of his engineering design approach, never more evident than five years earlier at the proposal review of an Army UAV in Great River; in that review, he displayed an unbending honesty to brief it as he saw it, coupled with the moral courage to hold fast in his position in the face of other influences from senior supervisors. Frauenberger's two principal strengths were a well-grounded systems engineering talent, and a gift—almost an artfulness—for approaching conceptual and preliminary design problems from a nonconformist viewpoint that opened up new possibilities. By now highly regarded and in demand for his ability to drop in and steer troubled programs through systems engineering difficulties, he found himself redirected from inside ATDC to track an emerging requirement from the Navy maritime patrol community for a replacement aircraft for the P-3C Orion. His instructions were to assess the program for its long-term viability and those of the potential competitors for their relative strengths and weaknesses, and to discern possible entry points for Northrop Grumman ATDC that might render irrelevant the obvious lack of a big patrol plane background.

Lapins and Frauenberger, both having had fairly long tenure in the former Grumman "iron works" (so called for its storied reputation for building rugged Navy carrier fighter aircraft), were part of the new Northrop Grumman for six years before coming into everyday contact on the same program. Not until late 1999 did they began working more closely together as events came to a head in the maritime patrol community's search for its P-3C replacement. Each individual contributed his personal experiences and talents in just the first phase—birth to adoption of the concept—of BAMS, which then would march through successive phases over the following fifteen or more years until an operationally representative BAMS (Triton) was flying. Lapins and Frauenberger were sought out for their knowledge and understanding of the concept and its potential employment, and both would return in following phases to keep BAMS on course.

An expression popularized by President John F. Kennedy after the Bay of Pigs posited that "success has a thousand fathers, but failure is an orphan." Any characterization of BAMS as a success for the Navy, for Northrop Grumman, or for any individuals involved must account for different phases of the program, the wax and wane of several acquisition and campaign strategies over that long fifteen-year march from inception to squadron stand-up. For the birth of BAMS, Lapins and Frauenberger were the midwives, with Ernest Snowden's later arrival to the team to add a deeper appreciation of the Navy customer and to construct a targeted market socialization campaign, working in concert with uniformed counterparts Easterling and Turner. Others played significant, often key, roles at this phase, and their contribution cannot be diminished—notably several individuals from the former Ryan Aeronautical company. Ryan would not join the Northrop Grumman amalgamation officially until the late spring of 1999, when Navy efforts to launch a P-3C replacement program with a large manned aircraft were already under way.

If there was a lesson to be drawn from the Northrop and Grumman merger experience in 1994, it was that cultural differences will exist and must be given due consideration in the pre-merger planning and post-merger integration. At the formation of the Northrop Grumman enterprise, recent corporate merger precedents that might inform a smoother integration of company cultures were not in ready supply. Northrop was in the vanguard of a wave of mergers fomented by Secretary Aspin's Last Supper, attempting to blend two longstanding aerospace giants that had each been in the defense marketplace for more than fifty years. Northrop seemingly paid little attention to any deep-seated cultural differences that may have existed between East Coast Grumman and West Coast Northrop, focusing instead on shared skill sets, overlapping and excess infrastructure, value to be captured from cost synergies, and consolidation of strategic planning and growth projection. Culture probably seemed a vague notion that was hard to deal with in an analytical, systematic, and practical way. Failure to ameliorate those cultural differences upfront perpetuated discord up and down the combined organization, which soon—and for several years—would have a negative influence on the execution of BAMS capture strategy.

By the late 1990s large mergers had become more commonplace both inside and outside the aerospace and defense segment, producing some good models worth emulating and some less than successful examples that suffered from debilitating culture clashes. For example, Daimler and Chrysler came together in the late 1990s as a merger of equals, but within a few years it was considered a fiasco. The German culture became dominant, and employee satisfaction levels at Chrysler dropped off the map. One unhappy joke circulating at Chrysler at the time was, "How do you pronounce DaimlerChrysler? Daimler! The Chrysler is silent."[28]

At the end of the 1990s the disparity in merger outcomes prompted several industry studies. Analysts at the Boston Consulting Group brought forth practices for coming to grips with culture in the post-merger environment that included an essential "cultural beliefs audit" as an element of due diligence. The audit was to be done well before the deal was closed to identify desired behaviors in both companies, to retain those deemed essential, and to recast process, directives, and even organizational structure for the new organization based on those adopted culture beliefs. "When cultural differences are not tackled from the start—overtly and with a clear plan—employees will cling to their legacy-company identity."[29] A necessary step implied by the study was the blending of the top level of leadership in the new company. A cursory look at Northrop Grumman's financial statements as early as 1995 indicate that former Grumman executives were nowhere to be found on the Northrop Grumman board or at the senior executive level under Kent Kresa.

However, Kresa and his team were learning as they grew. A stream of high-value acquisitions that included Newport News Shipbuilding followed the Grumman takeover. As an example of a more deliberate consideration of culture, it was written into the agreement with Newport News that Northrop Grumman would not repaint a huge shipyard crane that bore the company's former initials, *NNS*. Yet the culture issues left unaddressed through the Grumman acquisition would linger, smolder, and erupt at critical moments on the road to capturing BAMS.

CHAPTER 4
MOMENTUM
A REQUIREMENT EMERGES

For most of the last 40 years—and almost all of my career—the war that we focused on, that conflict that we were so concerned about, would come out of the Cold War.... Just ten years ago, we used to worry about Soviet submarines off the coast of the United States, just off of Norfolk, that could launch missiles that could strike Washington in eight or nine minutes time.... That was the guiding principle, the guiding assumptions relating to this kind of a war for most of the last four decades. That's all now gone.

—Gen. COLIN POWELL[1]

*B*lack Hawk Down, Mark Bowden's retelling of the disastrous 1993 incident in Somalia, pointed to a shift that had already begun in the U.S. Navy's maritime patrol mission:

Flying about a thousand feet over the helicopter was the Navy Orion spy plane, which had surveillance cameras that gave them a clear picture of the convoy's predicament. But the Orion pilots were handicapped. They were not allowed to communicate directly with the convoy. Their directions were relayed to the commander at the [Joint Operations Center], who would then . . . radio the command bird. Only then was the plane's advice relayed down to the convoy. . . . the Orion pilots would see a direct line to the crash site. They'd say, "Turn left." But by the time that instruction reached McKnight in the lead Humvee, he had passed the turn.[2]

This excerpt shows the turn away from antisubmarine warfare toward missions geared to above-water surveillance, with aircraft capability modifications to the basic P-3C Orion antisubmarine warfare version optimized now for anti–surface warfare and overland reconnaissance

> For the foreseeable future the continued fragmentation of the former Soviet state and its
> conventional armed forces have altered so fundamentally the character of the residual threat
> as to eliminate the capacity to wage conventional global war.
>
> —DALE VESSER, "FY94–99 Defense Planning Guidance,
> Memorandum to the Service Secretaries," February 18, 1992

in a derivative configuration named EP-3 Aries. Four years earlier, the Berlin Wall came tumbling down in a most visible manifestation of the Soviet Union's demise, and "the end of the Cold War brought a . . . virtual halt to Soviet submarine construction . . . and a massive cutback in naval operations."[3]

The U.S. Navy's maritime patrol force, dedicated for almost fifty years to airborne antisubmarine warfare surveillance and attack of Soviet submarines was, in 1993, stretching into new missions for the sake of its own survival. Deep reductions in aircraft numbers and community personnel were already under way, with inventory falling by 25 percent from 1990 through 1993 and another 7 percent within the following year.[4]

Although many of the aircraft shed from inventory were earlier model P-3As and P-3Bs, fatigue and service life projections for the remaining P-3Cs augured the need for a new aircraft development program to get expeditiously approved and under way. That need was now lead-time away, even with careful husbanding of remaining P-3C service life. The last new production P-3C came off the line at Lockheed's Marietta plant in 1990, so from that point forward, the focus of the maritime patrol community managers was on nurturing the next-generation maritime patrol and reconnaissance aircraft development, sustaining the structural viability of the P-3C fleet, and, with the emerging new emphasis on surface reconnaissance, introducing capability upgrades through Orion modernization for anti–surface warfare and reconnaissance. The reputed replacement aircraft for the P-3C—the much-anticipated next-generation aircraft—got an earlier start in the late 1980s when Lockheed bested other commercial derivative proposals with their P-3 redesign, coming in as the lowest cost and least risky technical approach. What evolved from Lockheed's winning concept was the P-7, essentially a warmed-over P-3C design, soon thereafter to be cancelled in 1991 for unbearable delay and cost overrun.

At that point, with no other workable option waiting in the wings, the Navy bore down on mission capability upgrades and service life extension of the P-3C to fill the gap. The steep decline in inventory over these years ameliorated the pressing service life issue, but only slightly. An inescapable fact of life as the new aircraft proponents watched that option slip further into an undefined future was that, for those increasingly rare occasions when the primary mission reverted to airborne antisubmarine warfare, a multi-engine turboprop aircraft was probably unbeatable in low-altitude submarine attack. For now, operation of existing P-3Cs would continue, albeit with the necessary modernization to heighten their relevance in surface surveillance. Lockheed seemed caught in a moment of indecision about how to best position itself for low-cost competition to earn the right to upgrade the P-3C to keep it relevant. The company had fumbled a golden opportunity to preserve and extend its Orion franchise that would not come around again.

To remake the existing P-3C and give it new utility to take on the anti–surface warfare mission, the maritime patrol community embraced the anti–surface warfare improvement program as the premier upgrade path for the inventory. This improvement program consisted of new and improved onboard avionics to optimize the aircraft's performance and survivability in the anti–surface warfare mission. With it came a much-improved "standoff, Over-The-Horizon Targeting capacity, and interoperability with Command, Control, Communications and Intelligence networks, and access to pertinent real-time tactical intelligence."[5] Perhaps the most distinguishing feature of the upgrade was the addition of a new inverted synthetic aperture radar that used target movement to generate a two-dimensional high-resolution image, ideal for differentiating superstructure outlines that could help with surface ship classification. Augmenting the radar was an electro-optical system for long-range ship identification once cued by detection and classification from the radar.

Importantly, as the maritime patrol and reconnaissance community would come to realize more than a decade on, capabilities enhanced by this radar and the companion electro-optic/infrared sensor upgrades can render similar performance from a long-endurance unmanned aircraft as one encumbered by the freight of a dozen aircrew. Yet only P-3Cs with the most advanced avionics configuration to that point were to receive the improvement program updates, limiting its installation to less than half the inventory. Finally, to address the inexorable decline in service life, the community undertook a sustained readiness program to focus on areas of high corrosion—the most pernicious degrader—by reconditioning or replacing those areas that were beyond the capability of squadron or intermediate level repair. Conceived in 1993 but not implemented until the following year, sustained readiness "encompassed the unprecedented disassembling of the aircraft, dividing the airframe up into four major components: empennage, wings, engine nacelles, and main-tube fuselage, with each component subsystem refurbished separately."[6]

Lockheed failed to tenaciously protect its franchise by allowing the sustained readiness contract to be awarded to Raytheon in Greenville, Texas. Raytheon pursued a program strategy that predetermined a set of "core" replacement parts to be installed on every inducted aircraft and a set of "conditional" parts that would be used if, upon inspection of the disassembled aircraft, their installation was indicated. From the first induction and disassembly onward, corrosion proved much worse than anticipated: "The Navy chose to induct their most severely corroded airframes into Sustained Readiness, which overwhelmed the program. A stop work order was issued in 2000 after the completion of only 13 aircraft. Eighteen of the remaining 19 aircraft that were inducted into Sustained Readiness were ferried to Lockheed Martin's Greenville, South Carolina, facility to receive a modified version of Sustained Readiness, while the last one completed repairs in Texas."[7]

Raytheon's executive in charge of the sustained readiness program, Phil Teel, found the going tough for continued career growth in Greenville after termination of the program. In 2000 he would land nearby at Northrop Grumman's integrated systems sector headquarters in Las Colinas, Texas, chastened somewhat by his sustained readiness program experience but buoyed by the prospect of new opportunity to regain an upward career trajectory. In short order, sector head Ralph Crosby assigned him to run Northrop Grumman's Bethpage site, where he would oversee an emerging BAMS program that was already getting some notice in the Navy maritime patrol program management office. Delay in recapitalization of the P-3C force, worsened by collapse of the sustained readiness program and further diminished by the decline of its historic and principal threat, had the effect of making maritime patrol more costly to maintain while adding inflationary pressure to what was becoming a very large outlay required to ramp up new aircraft development. Fortuitously, by the end of the 1990s, conditions in maritime patrol were beginning to favor more creative, divergent thinking about how best to recapitalize the force, which then stood at just over 250 operational aircraft.

Other consequential influences began to converge in 1993 that would, by forethought or happenstance, help to shape a more conducive environment for BAMS to take hold in Navy thinking. A newly installed administration under Bill Clinton, confronting a much-reduced Soviet threat, was moved by the tantalizing prospect of a peace dividend and the defense investment that might be salvaged as a result. In March 1993 newly confirmed secretary of defense Les Aspin announced a "whopping cut to the defense budget. Incredibly, he made his cut without calculating the impact the reduction would have on force capability. . . . Details would be worked out in a Bottom-Up Review to follow. It was an exercise to justify a blind budget cut, but it shaped the force for a decade."[8] Although he acknowledged the deep cuts already made by the preceding administration, Aspin proposed to double the cumulative reductions since 1990 and set the future years defense plan at $245.2 billion below a previous agreement between the George W. Bush administration and Congress, noting that "this budget begins to use resources freed by the end of the Cold War to help at home."[9] Armed Services Committee chairman Senator Sam Nunn was appalled, commenting: "We have been dealing with numbers grabbed out of the air; no one knows where these cuts are going to come from."[10]

To camouflage the depth of these new cuts, the fig leaf of a new remodeled force structure rationale was required of the joint staff, which dutifully assembled the "two nearly simultaneous major regional contingencies" force level to conform to the new budget numbers. The cuts were codified four years later by the Quadrennial Defense Review (QDR). "The mismatch between strategy and resources persisted through the 1990s. . . . The defense budget did not bottom out until 1998, by which time it had been cut for thirteen years in a row. Readiness rates were down. Older equipment wore out and was not replaced."[11]

For the maritime patrol and reconnaissance community, sitting at the lower end of aviation funding hierarchy throughout the 1990s, the upshot was clear: P-3C readiness suffered from a dearth of funding for maintenance and repair, while P-3C modernization for mission expansion limped along at a halting and inefficient pace. In 1993, however, recapitalizing with a new-start aircraft replacement program would remain wishful thinking for the foreseeable future. It would be another six years before momentum could be regenerated for the new aircraft. Yet the community had a P-3C inventory, over half of which were built in the late 1960s, that would have reached the end of its useful life by 2000. The community regrouped around a service life assessment program, initiated in 1996, that centered on a full-scale fatigue test to determine precisely the fatigue life remaining and "structural inspections, modifications, replacements and redesigns necessary to sustain the P-3 inventory until at least the year 2015."[12]

A COMMUNITY REDEFINES ITS MISSION

As the service contemplated a service life extension program informed by the service life assessment results, the maritime patrol program office was urgently looking ahead to the introduction of a replacement, the multimission maritime aircraft. The planning timeline was bounded by the recent abysmal failure of the sustained readiness program at the front end and the uncertain outcome and cost of a service life extension at the back end, with the remaining fatigue life of the inventory—and with it, fleet availability much beyond 2010—still a moving target. The transition out of the P-3C to new equipment was well past the lead time required to successfully navigate the defense acquisition maze for a new-design large patrol aircraft. The maritime patrol program office continued planning for service life extension but by 1999 began laying the groundwork for entering the maze with MMA preliminaries.

Capt. Brendan Gray, overseeing the program management office's MMA effort, established a coordination team staffed by Navy captains from various fleet areas, principally Second Fleet from the East Coast and

Third Fleet from the West Coast, and a fleet user team composed of more junior P-3C operators at the lieutenant and lieutenant commander grade who engaged in a the still-novel process of quality functional deployment to define baseline requirements for the new aircraft. The process was thought to have originated in Japan several years before as a method for translating the "voice of the customer" into engineering characteristics. Still fairly new at the end of the 1990s in Navy acquisition, quality functional deployment proved a usable mechanism for ranking warfighter preferences against system attributes to forge a consolidated operators' best sense of the qualities needed to perform a range of missions.

These tentative steps toward finding consensus on MMA performance were useful and a bright portent of real progress toward a P-3C replacement. Further delay would have been anathema to future viability of the maritime patrol force. For most of the 1990s, operational tempo had not slackened in the least, and P-3C inventory service life was grinding down inexorably. In only five years, from 1991 to 1996, the force shrank by half its numbers— from twenty-four active-duty squadrons with twelve reserve squadrons to twelve active and six reserve squadrons. If there were any doubts among fleet commanders, Pentagon force planners, or even older hands in the patrol force that VP squadrons had made a wholesale shift from over-water antisubmarine warfare to overland collection of ISR, a succession of global operational deployments throughout the 1990s was evidence enough.

An illustrative experience was that of a junior officer newly arriving in VP-11 from the P-3 replacement training squadron in 1993. Lieutenant (junior grade) Dennis Hayden was slotted for the usual aircrew position taken by an incoming junior naval flight officer, that of navigator/communicator. Within months of checking aboard VP-11, Hayden and his entire squadron deployed to Roosevelt Roads in Puerto Rico to preposition in range to support counternarcotics and illegal immigration flights in the Caribbean. Deployment included not just eight P-3Cs and their aircrew, but all of the nonflying infrastructure and personnel as well. Further detachments of two aircraft relocated from Roosevelt Roads to more austere sites in Panama, Honduras, or Curacao to close the range gap to the operating area and enable more on-station time. Four to five times a week, an aircrew would launch on a ten- to twelve-hour surface search of a rectangular box, typically 150 by 300 miles, at 1,500 feet, most limited to the fairly inefficient APS-115 radar, electronic support measures, and visual means to detect, track, and report targets to a task force command center at Key West, Florida. Hayden would establish and maintain contact with their home base over high-frequency communications with a tactical support center in Roosevelt Roads. Near-constant communications were maintained via a high-frequency set, with some ultra-high-frequency satellite communications available in several squadron aircraft. In-area Coast Guard and Navy surface ships could be reached on ultra- or very-high-frequency bands. However, since range and on-station time limits were an issue most days, a ready alert one aircraft would be on call at Roosevelt Roads for surface targets of interest, able to get airborne in one hour and fly out on a vector taken from an Air Force airborne warning and control system, deployed Navy E-2C, and a Customs and Border Patrol air search–capable P-3. A ready alert two would be on standby for air targets, able to launch in thirty minutes to fly out on a vectored heading for intercept of the target.

Within a year of the Caribbean mission, Hayden—now elevated to tactical coordinator as a full lieutenant—and the squadron again deployed, this time to Sigonella, Italy, for operations in Eastern Europe as part of Operation Decisive Edge. Patrolling the Adriatic, they typically flew at 20,000 feet to stay above lethal range of deadly shoulder-mounted surface-to-air missiles, while displaying an illuminated radar and communications signature to light up potentially hostile sensors on the water and over land for friendly intelligence monitors. Decisive Edge ran concurrently with Operation Joint Endeavor in response to the Dayton peace accord, with VP-11 now supporting the Allied Implementation Force on the ground in and around Bosnia. For the squadron, missions moved increasingly over land with more difficult mission geometry to skirt unauthorized airspace, with a one-way commute from the operating base in Sigonella of more than 450 miles.

To prosecute a preassigned target folder, much depended on successful employment of onboard electro-optical long-range high-resolution cameras and infrared turret-to-image surface infrastructure—e.g., an automobile plant and various truck and fuel farms—the use of which was proscribed

A U.S. Navy P-3 Orion Surveillance plane captured imagery of a standoff between Bosnian Muslims and Russian soldiers in the town of Jusici, according to service officials here. The incident occurred when Bosnian Muslim refugees seeking to return to their homes in the northeastern Bosnian town—within the zone of separation—faced off with Russian soldiers after it was determined some of the refugees were carrying weapons in violation of the Dayton Peace Accords. The P-3, which was designed for anti-submarine warfare (ASW), monitored events on the ground. The P-3 was outfitted with antennas derived from the Navy-Marine Corps Pioneer unmanned aerial vehicle (UAV) program.

—BRYAN BENDER, "Navy Surveillance Plane Monitored Standoff in Bosnian Town," *Defense Daily*, October 10, 1996

by the terms of the accord. From Sicily, VP-11 redeployed a detachment to Western Africa as part of Operation Assured Response to provide overhead watch on evacuation operations in Monrovia, Liberia. Political and tribal instability was threatening the U.S. embassy and its staff, and the P-3C employed its suite of radar and electro-optic/infrared sensors to monitor the approaches to the embassy and coordinate ground and other aircraft as evacuations got under way. The P-3Cs of VP-11 would relieve each other on-station after commuting more than seven hundred miles one way from their operating base in Senegal. In five years with VP-11, performing duties essential to the P-3C missions, Hayden never dropped a sonobuoy to prosecute an uncooperative submarine contact in real-world scenarios.

To the extent that VP-11's experience was representative of P-3C deployments generally throughout the 1990s, the maritime patrol force showed great alacrity and adaptability in redefining its primary mission as surface surveillance and reconnaissance, and occasionally targeting, when the Soviet submarine threat receded. In truth, many, if not most, of VP-11's deployments exemplified the utility of the P-3C in ISR, and missions around the globe would have been considerably more difficult for U.S. fleet and area commanders without the services of the P-3s, which in this role "spent tens of thousands of flight hours providing intelligence, surveillance, and reconnaissance support to Operations Desert Fox, Provide Promise, Deliberate Force, Allied Force, Enduring Freedom, Iraqi Freedom, and Odyssey Dawn."[13]

Getting to the operating areas involved lengthy commutes from a deployment base and demanded long-duration search patterns over pre-briefed surface features. To make this work, large four-engine airplanes consumed considerable fuel while transporting ten or eleven aircrew drawing regular pay and flight pay allowances, not to mention the cost associated with major deployments that dispatched many more nonflying personnel and their equipment to far-flung mobilization bases. The P-3C filled a niche that could not have been provided easily by other means, so operations and support costs were not a front-burner issue when real-world scenarios dictated the need. Only as the program office began to assign notional cost to quality functional deployment results did a picture begin to emerge that suggested replacement costs would dramatically exceed MMA's future share of naval aviation obligation authority. Continuance of manned airborne antisubmarine warfare (ASW) was seen as inviolable: "By 1999, P-3s were better known to joint warfighters as good overland ISR platforms than sub hunters. . . . This tasking did little to hone aircrew ASW skills."[14] The vagaries of submarine search and attack would still require direct human involvement and reaction in the blur of airborne localization and torpedo attack. Open-ocean, long-duration surface surveillance was not a particularly economical chore for a crew of ten in an aircraft optimized for low-altitude ASW. At this moment, however, a viable alternative was not on anyone's radar. That would soon change.

NEW LIFE FOR RYAN AERONAUTICAL COMPANY

The defense industry consolidation unleashed by the Last Supper continued almost unabated for another five years. At Northrop Grumman, Kent Kresa's appetite for aggregation of pedigreed defense firms was not sated with the acquisition of Grumman and Vought. Despite almost being consumed itself by Lockheed Martin in the industry's acquisition frenzy in 1998, Northrop Grumman resumed its march toward growth by agglomeration, with the merger with Lockheed Martin having been voided by the U.S. government apparently in response to aggressive opposition from Raytheon and its political supporters. Throughout the 1990s, Kresa's pursuit of companies that brought additive value to his strategic vision for Northrop Grumman led to purchases of Westinghouse Electric

Systems to form Northrop Grumman's electronic systems sector and of Logicon to form his mission systems sector. The jewel in the crown for the aerospace business, however, was Ryan Aeronautical, acquired from Teledyne in 1999 for $140 million. Kresa stated that the Ryan acquisition was "an excellent strategic fit with many of Northrop Grumman's business areas and strengthens our surveillance and precision strike capabilities."[15]

Until this acquisition, Ryan Aeronautical was really a fringe player, well positioned in the so-called revolution in military affairs among a constellation of companies with unmanned systems but lacking the critical mass to carry the program beyond development into serial production and to expand its application to different services and different missions. The Ryan deal was a benchmark for Kresa's aggressive move to reshape Northrop Grumman "from a dwindling builder of combat aircraft to an electronics power house that specializes in growing surveillance and precision strike markets."[16] Important acquisitions would follow: Litton Industries in 2001, adding depth in information systems, marine electronics and navigation, radar systems, and shipbuilding with the former Ingalls shipyard; Newport News Shipbuilding that same year, adding the single producer of the nation's aircraft carriers; and, most significantly, TRW in 2002, adding laser weapons and a healthy portfolio of communications and classified reconnaissance satellites.

None of these later acquisitions were to have as much influence, even indirectly, on BAMS as the acquisition of diminutive PRB Associates in Patuxent River, Maryland, in 2000. The company, which employed only four hundred people outside the gate at the naval air station, came to Northrop Grumman's attention by way of the former Grumman staff in Bethpage. That staff then still supported the electronic attack mission for the Navy and Marine Corps through the EA-6B Prowler product line and was familiar with PRB's niche service in scene generation for computer simulations and mission planning for the EA-6B aircraft. PRB's skills meshed nicely with Bethpage's electronic attack franchise. Much more important for BAMS, however, was PRB's incumbent position as prime support contractor to the Navy's lead agency for information technology and communication architectures, the Space and Naval Warfare Systems Command. PRB's support expertise was retained for tactical support centers, resident in P-3C operating base as the operations planning cell, and mobile operations command centers, the active control stations for P-3C missions. These two provided somewhat the same capability, but the mobile operations centers could pack up and reposition in temporary habitats to support P-3 operations at more remote, austere deployment sites. These were essentially the mission nerve centers for P-3 maritime patrol operations. More than a dozen tactical support centers occupied brick-and-mortar installations in proximity to P-3 operations. With only a little imagination, these tactical support centers and mobile operations command centers could be reconfigured as ground infrastructure to host an increased flow of intelligence information from a maritime patrol unmanned aircraft system. They seemed to be the logical node for accommodating something like the two-person mission control element that then managed Global Hawk flight planning and sensor positioning in joint exercises for the Air Force developers. PRB's role, though not dominant in the BAMS campaign, would emerge later as a key enabling factor.

Meanwhile, tiny Ryan Aeronautical would experience growing pains integrating into the much larger Northrop Grumman, not unlike Grumman's assimilation discomfiture five years before. A Ryan engineer commented at the time of their takeover: "We used to have 300 people total on the program [Global Hawk] and we all worked at one facility in San Diego. Northrop came in and assigned 3,000 people to the project who were located at multiple facilities around the country."[17] Watching from afar, former Grumman employees abused by autocratic Northrop management practices in the earlier acquisition had their own assessment: "All that was innovative, adaptable, and responsive about Ryan will now be crushed by the weight of the bureaucratic behemoth, Northrop." Both fears at Ryan and wry judgments at Grumman were overstated. What Northrop Grumman leaders wisely did this time was to conduct program and technology exchanges between the leading engineers at the company's various development centers with Ryan in San Diego. Included in these visits were business development and program management people from around Northrop Grumman to encourage widest understanding of Global Hawk development, capability, and future plans—and to offer helpful suggestions based on their own customer experiences.

In the first wave of visits, key concept developers from Bethpage and from Northrop Grumman's Washington office visited Ryan Aeronautical, still lying athwart Lindbergh Field in San Diego. These visits, and the knowledge and personal introductions obtained thereby, would prove to be of inestimable value in setting the conditions for successfully migrating the Global Hawk concept toward a Navy BAMS application. If a football arena was the metaphor for where this transition would play out, then the contest was defined by several defensive linemen that came on as blitzing challenges, which, each in turn, would have to be stood up and pushed back for the offense to move the ball down the field:

- fleet opposition to unmanned aircraft viewed as detrimental to manned aircraft—the challenge of ingrained culture

- the maritime patrol community's own playbook—its accumulated performance in airborne ISR in the absence of an antisubmarine warfare threat, which suggested to them that the mission had been, and could continue to be, performed best with a manned crew (an attitude that theoretically could be turned to advantage by an unmanned system that actually performed most of the airborne ISR mission more effectively and at a fraction of the cost)

- the internal resistance from the former Ryan Global Hawk team that, most likely due to their tenuous newness and parochialism, sensed that Bethpage might be encroaching on their fiefdom

- the industry primes who had thus far shaped for themselves a preferred position in the emerging MMA competition with large aircraft candidate concepts—their position would be undermined by an unmanned aircraft if it showed any sign of gaining traction with the Navy.

The contest would have to be competed in observance of the rules of the game set by the federal acquisition regulations and directives. And the players themselves were now walking on to the field:

- Ryan—and its Global Hawk team—were integral parts of Northrop Grumman with a commission to exploit Northrop Grumman's far-flung resources to further advance the program.

◆ PRB was on board to address the ground control infrastructure that then served the maritime patrol community and could host the command and control function for the unmanned system.

◆ Key people from the Bethpage technology center—Lapins and Frauenberger—were fully immersed in the intricate machinations of the program.

◆ At that moment, Ernie Snowden was drafted into the team—not by direction from company officers in Bethpage or sector headquarters, but simply because Bethpage was an internal customer served by whatever strategy and customer perspective he could bring to the effort from a position in the Northrop Grumman Washington office.

In Washington and in Patuxent River, on the customer side of the equation, the principal disrupters, Captain Easterling in the maritime patrol program office and Rear Admiral Kilcline and Lieutenant Commander Turner on the CNO's staff in the Ranch, were in place or on their way to their assignments where each would add critical advocacy, activism, and promotion of the BAMS concept when the moment demanded. The actions of these individuals in concert would ignite an idea and nurture it through its early difficult infancy.

The call to action was the opening presented by the MMA initiative. The maritime patrol community, having rekindled purpose and efficacy in airborne ISR, was long past the desired start date for a replacement aircraft. To perform adequately in ISR and retain fallback capability in

---

The U.S. Navy this year will begin its analysis of alternatives for the land-based Multi-mission Maritime Aircraft planned to replace the Lockheed P-3, EP-3, E-6, and Navy C-130s. The three options under consideration are a new aircraft design; a derivative of the Boeing 737, Lockheed C-130J or other existing platform; or remanufactured P-3s with a new electronic suite. P-3 remanufacture is low risk, as the service is undertaking a detailed life extension analysis and will know exactly what shape the aircraft is in, according to a Navy official. Although MMA is supposed to enter service around 2010, the service has budgeted only for studies.

—PAUL PROCTOR, "New Navy Aircraft Studied,"
Aviation Week & Space Technology, June 28, 1999

---

antisubmarine warfare for a resurgent submarine threat, a new MMA would necessarily be a compromise of design and mission—and would demand an inordinate share of naval aviation procurement dollars if it replaced the P-3 on a one-for-one inventory basis.

As surely as 1993 ushered in a synchronicity of geopolitical, acquisition, force structure, and technology events to shape an environment conducive to BAMS, the year 2000 bore witness to a new rollout of major changes to DOD acquisition policy, Navy staff organization, and fleet experimentation that proved both a drag and stimulus on the further evolution of BAMS. At the close of President Clinton's eight years in office, his third under secretary for acquisition signed a departmental instruction in October 2000 that introduced a number of refinements to the acquisition process for major systems. The most significant changes had to do with criteria and decision points for entry and exit of phases in the acquisition process continuum. If a program had passed its concept phase but had not progressed into engineering and manufacturing development, it now had to negotiate a systems development and demonstration phase instead. The earlier engineering and manufacturing development and new systems development and demonstration shared a focus on reducing risk and maturing engineering design; however, development and demonstration now imposed a greater burden on service acquisition executives, program executives, and program managers to treat cost as system requirement, essentially an independent variable in the program cost calculation; inject competition at every phase, which imposed a further burden to seek open architecture systems to permit unfettered competition; and establish key performance parameters as a gauge of system utility, wherein each parameter would be assigned a threshold value (defined as minimum acceptable) and objective value (defined as most desired by the user), with each parameter traceable to an operational requirements document and further back to the original mission needs statement. BAMS would be just exiting its concept study phase as the new instruction was codified, and its strictures would bear directly on the timeline and shape of its program evolution.

This acquisition framework (see figure) displays the procedural steps required by DOD to mature a major defense system from initial concept

Defense Acquisition Management Framework

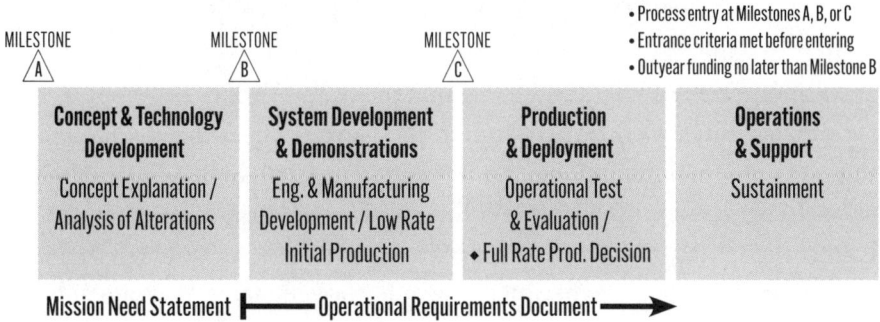

- Process entry at Milestones A, B, or C
- Entrance criteria met before entering
- Outyear funding no later than Milestone B

MILESTONE A	MILESTONE B	MILESTONE C	
Concept & Technology Development	**System Development & Demonstrations**	**Production & Deployment**	**Operations & Support**
Concept Explanation / Analysis of Alterations	Eng. & Manufacturing Development / Low Rate Initial Production	Operational Test & Evaluation / ◆ Full Rate Prod. Decision	Sustainment

Mission Need Statement ┣━━━━━ Operational Requirements Document ━━━➤

Source: Department of Defense Instruction 5000.2 "Operation of the Defense Acquisition System" dated October 23, 2000.

through production and post-production logistics support. The framework would be superseded by a revised instruction issued by the Bush administration's incoming deputy secretary. The revised guidance further divided the acquisition phases and added more decision points but essentially did not change the structure of the framework. For our narrative, early strategy planning for BAMS was linked to the October 2000 instruction and underwent only minor adjustments to accommodate the revised framework in May 2003.

As the 1990s ended, the influence of CNO Adm. Jay Johnson's staff organization—charged with resources and requirements—was waning but still held sway over the Navy planning, programming, and budgeting buildup. This work was increasingly dominated by simulation and scenario modeling, but when Adm. Vern Clark succeeded Johnson as CNO, the platform requirements oversight was stripped from one staff office and reassigned to another. The so-called barons, with unquestioned ownership of aircraft, submarine, and surface combatant requirements, now held sway. The fortunes of BAMS would be at the discretion of the directorate for air warfare, even though at its creation, it lacked a branch that was recognized as the advocate for unmanned systems. That would soon change.

THE DISRUPTERS ADD DEPTH TO THEIR BENCH

Before retiring, Admiral Johnson made further organizational changes that would bode opportunity for BAMS, if only tangentially. He elevated the Naval War College presidency to a three-star billet and placed Vice Adm. Art Cebrowski in that position where he could "personally proselytize on aspects of new Navy thinking, e.g., network-centric warfare and unmanned systems." And he created the Navy Warfare Development Command under Cebrowski's lead at the Naval War College, charging that new command with concept development to revitalize Navy operational and tactical thinking. Within the Navy Warfare Development Command, a new maritime battle center was created to run fleet battle experiments.[18] These experiments would occasionally thereafter be derided throughout the fleet as no more than demonstrations or stunts, but they would prove to be an innovative and often effective way to highlight new systems and technologies and their potential warfighting utility in near-real-world scenarios while immersed with real fleet units. Fleet battle experiment Juliet would become a first target of opportunity in a BAMS socialization campaign.

In his first year as CNO, Admiral Clark would move the warfare development center out from under the Naval War College to a position subordinate to the commander of U.S. Fleet Forces Command in Norfolk, Virginia, to tie its activities closer to fleet requirements. Admiral Clark would also carve out a special cell for preparing the Navy response to the second congressionally mandated Quadrennial Defense Review under Rear Adm. Joe Sestak and his QDR planning cell, which was then moved under Vice Adm. Ed Giambastiani. Sestak soon was able to take his QDR briefing on a road show of informational presentations to Navy flag officers and fleet commands, wherein the first images of an unmanned aircraft resembling a Global Hawk would appear. It was not by accident that Global Hawk could be clearly discerned in the briefing slides, and Northrop Grumman had a direct role in influencing the look and feel of Sestak's briefing. It came about largely due to the influence of a new face in the Northrop Grumman Washington office, a transplant from the assimilated Vought Aero Products Washington office.

Arriving at Vought a decade before from a staff position in the Office of the Secretary of Defense, Ernie Snowden was not widely known in the Northrop Grumman divisions outside of Washington. His experience as a newly arrived Washington office staffer derived in part from prior work in the Office of the Secretary of Defense, where he routinely deconstructed and identified issues within service budget request submissions when not authoring congressional testimony for the political appointee he directly served. Even before arriving as a Pentagon staffer, Snowden held a significant position in NAVAIR as the budgeteer and money manager for the command's operations and maintenance funds. Snowden had been schooled in the intricacies of the Pentagon's acquisition processes, in the rules governing budgeting and obligation of the different colors of defense monies, and in the personalities and their offices that pulled the levers behind the curtain to make defense acquisitions go. Vought made him the sole company representative to the Navy's Pentagon and Naval Air Systems Command staffs from their Washington, DC, office. When Vought entered a teaming arrangement with partners Northrop and Grumman for the Navy advanced tactical aircraft (ATA) competition for a new penetrating Navy attack platform, Snowden was dispatched to a covert location on the West Coast as the Vought business development co-lead for campaign strategy. When the team's ATA bid was withdrawn, he rolled directly to a follow-on detail as the capture campaign co-leader on the Vought–McDonnell Douglas team competing for the Navy A-X advanced stealthy strike aircraft.

Vought's acquisition by Northrop, consummated in 1994, cast Snowden's aerospace and defense industry future into doubt. It was clear that Vought's diminished capacity would likely not require additional representation in the newly formed Northrop Grumman. A layoff notice seemed to be in the offing. In a few weeks, when the newly consolidated Northrop Grumman next partnered with Lockheed Martin's joint strike fighter team, Snowden went to work carving out a meaningful role for himself in the new Northrop Grumman. He lobbied senior Navy acquaintances as well as contacts in the Washington offices of Grumman and Lockheed to sponsor his reassignment, in part due to his recent ATA and A-X experience, as the sole Navy requirements representative on Lockheed Martin's joint strike fighter team. He would work as part of a three-member customer requirements cadre with Lockheed's Air Force and Marine Corps representatives.

Entering the late 1990s, Snowden found himself drawn closer to the Bethpage operation, then known as the airborne early warning and electronic warfare systems division of the Northrop Grumman integrated systems sector, where he assembled and put in motion congressional lobbying campaigns to generate plus-up appropriations for E-2C radar development and EA-6B life-extending structural modifications. Such a nonlinear career progression gave him a breadth of experience with the defense budget process, Navy requirements development, congressional appropriations processes, personalities and organizational behavioral norms in Congress, the CNO's staff, and NAVAIR—all of which would be enormously applicable to the gestation of the BAMS program. When he got his first glimpse of the Navy's QDR outline and initial organization under Rear Admiral Sestak, that accumulated experience told him that it lacked any association with or even passing reference to the broader trends then emerging from the Center for Naval Analyses, the CNO's strategic studies group in Newport, the CNO's warfare integration division staff in the Pentagon, or even less Navy-centric defense policy writings in the public square.

Instead, Sestak's group was driving their presentation back to a discussion of global choke points and the Navy's role in freedom of the seas, with no mention of the digital networking of disparate naval and joint forces to achieve precision effects in the littorals and beyond at the globe's hotspots. Through intra-office discussions and reports, Snowden became privy to company individuals and their efforts supporting work for the Air Force's own QDR preparation, and he saw that their work was embarrassingly far ahead of what the Navy was doing. He sought a private meeting with Sestak—only Sestak's deputy sat in—to lay out the major shortcomings and to offer behind-the-scenes assistance to correct the Navy's misdirected and unimaginative effort. Perhaps recognizing the need for a fresh approach, Sestak went along willingly, to the point of allowing Snowden to arrange and host several whiteboarding sessions for a combined session of Sestak's staff and Northrop Grumman staff in the Northrop Grumman Washington office. The attempt was to outline a radically different QDR package, even to the point of connecting Sestak's staff with an animation graphics company in Los Angeles that Snowden recommended for the most impactful visuals.

An intended benefit of this unsolicited help was the liberal reference to the Northrop Grumman stable of products then in development or in production. Sestak's eventual briefing was viewed in the most favorable light by the entire flag officer leadership in the Navy. Admiral Giambastiani sent kudos to the Northrop Grumman sector president and Washington operations head, citing Northrop Grumman's powerful support in general and Snowden's direct contribution to the Navy's QDR effort in particular. This experience and related episodes enhanced his credentials and value to what was emerging as the BAMS team.

Yet unlike Lapins and Frauenberger, Snowden was not a practicing engineer. He possessed sufficient engineering background that while working primarily as a business developer, he had become adept at using systems engineering logic to architect capture campaigns for new business opportunities, a method still rarely employed by his more recently hired business developer peers who had served a lifetime career in military aviation. Snowden's practice of building and executing a systems engineering–based marketing campaign won over a following of adherents. He had become practiced, within a corporate environment, at defining a desired endpoint—usually a strategic contract win—and building in the progression of milestones and alternate paths to the next gate leading to the end point, all while setting tactics for the intermediate steps that had to be met and relationships to be made to stay on the critical path. It was this strategic perspective for new business captures—and the knack for how to implement it—that ran through his body of work experiences. There would be a place for this skill set on the BAMS team alongside Lapins and Frauenberger.

CHAPTER 5

DISRUPTION
MOMENTUM STALLS

The culture of any large institution of long standing almost always mitigates against ready acceptance of new concepts or, in the case of the military, against new weapons systems. The Navy is not immune to the effects of this phenomenon.

— *Autonomous Vehicles in Support of Naval Operations*[1]

Harvard Business School educator Clayton Christensen describes disruptive innovation this way: a situation where incumbent businesses focus on high-end, high-priced solutions to the exclusion of other offerings in a market segment and are challenged by new market entrants that begin by "successfully targeting those overlooked segments, gaining a foothold by delivering more-suitable functionality—frequently at a lower price."[2] Christensen and his co-authors had in mind a private sector consumer market, but his definition could be easily applied to the appearance of an unmanned aircraft in the traditional mix of large multi-engine patrol planes vying for the Navy's MMA program.

The maritime patrol and reconnaissance community had every expectation that their MMA acquisition would be narrowly structured to deliver a manned aircraft that required only type training—essentially training limited to aircrew familiarity with flying qualities and maintenance differences for ground crews—to permit a fairly seamless transition from their P-3Cs. Interested and capable aerospace companies looked no further than designs that accommodated a large crew and featured like functionality, with some modernized appurtenances for easing man-machine interface and updating sensors. The disruption that unfolds from here is the wholesale reordering of force structure, manning, training, tactics, and acquisition, thanks to the injection of unmanned aircraft technology. It came about as a result of not only new innovative technology, but also the way in which the technology and its employment were conceived, socialized, and maneuvered to gain trust, promote advocacy, and obtain widespread community acceptance: "True innovation springs from a combination of a deep understanding of customers' needs and a willingness to approach a problem from a different angle—connecting the dots where no one else has."[3]

Both intimate customer familiarity and the completely unexpected one-off approach to the customer's problem were important pieces in the progression to BAMS. The manner in which a marinized Global Hawk became the technical innovation that disrupted conventional thinking was not immediately foreseeable to any but a handful of people who had the vision, patience, and persistence to build broader enthusiasm and gradually involve a wider circle of convinced like-minded advocates. It originated inside Bethpage's ATDC, but the MMA program had to show some stability and direction in order for it to emerge.

With the quality functional deployment (QFD) analysis at an end and MMA mission attributes narrowed and prioritized, the Navy coordination team and fleet user team were disbanded. Industry representatives had been invited to observe the QFD proceedings but were relegated to nonvoting participation from the sidelines. Incoming Cdr. Alan Easterling, eager to take ownership of the process and sustain the momentum, relieved Captain Gray at this point as the deputy for the maritime patrol program management office for MMA. Easterling was granted authority by Rear Adm. John Nathman, director of air warfare on the CNO's staff, to immediately invite a trade association, the National Defense Industrial Association (NDIA), to organize the means, media, and frequency of

---

In September 1999, the Navy completed a two-year MMA concept study that examined single-platform, manned air-vehicle alternatives to the P-3C and EP-3. In March 2000, NAVAIRSYSCOM received approval from the Office of the Secretary of Defense (OSD) to initiate the formal concept exploration. Guidelines are for nonrecurring manufacturing costs not to exceed $800 million and recurring costs not to exceed $55 million per aircraft (based upon a 251 aircraft buy).

—KENNETH SHERMAN, "P-3 Replacements Proposed,"
Journal of Electronic Defense, October 1, 2000

---

contact with interested industry parties to begin a preliminary but very critical look at MMA design concepts. He opened up the trade space for industry consideration of MMA as a replacement for not just the P-3C but also other large Navy aircraft that conceivably could host the maritime patrol mission, including the EP-3E Aries signals intelligence aircraft, the C-130 Hercules transport, and the E-6 Mercury, the converted Boeing 707 passenger aircraft that served as the strategic communicator with the Navy's ballistic missile submarine fleet.

The QFD work became the jumping-off point for gauging performance of several notional aircraft designs. Throughout this process, Howie Frauenberger, now assigned to Bethpage's ATDC, dutifully maintained his presence at every Navy coordination team and fleet user team meeting, at the ready to inject an opinion on aircraft design and performance consideration that might be overlooked by the fleet operators. Equally, at every NDIA-hosted meeting involving Navy participants and industry attendees, Frauenberger was watchful for any opening for potential industry teaming or direct contract support to the Navy for independent simulation or analysis, with an eye to uncovering a play for Northrop Grumman. Not considered a player with a front-runner aircraft candidate in the mix, Northrop Grumman's best option at the beginning of this NDIA process appeared to be some derivation—a growth, non-carrier-suitable version—of a design concept then under early development by the Navy called the common support aircraft (CSA). Another small group of Bethpage design experts collocated in Frauenberger's office suite was already working under a Navy contract to develop concepts for the CSA.

The Navy saw the CSA as a means to consolidate several carrier-based support aircraft into a single multimission platform for recapitalizing the E-2C airborne early warning aircraft, the C-2A carrier logistics resupply aircraft, the S-3A carrier-based antisubmarine warfare aircraft, and the ES-3B, a carrier-based version of the S-3A designed for signals intelligence. Each of these aircraft was a mid-sized cargo carrier in one sense, because each carried either cargo or a special outsized sensor package or multiple crewmembers. But each was limited in overall size because a key requirement was operating from the carrier flight deck. In truth, each was much like a smaller version of the MMA idea but intended for entirely different missions and aircraft carrier–capable. Northrop Grumman was the incumbent manufacturer of two of the CSA aircraft types under consideration (E-2C and C-2A), and its CSA trade studies were leading the Bethpage designers to a preferred solution based on a stretched and widened E-2C fuselage.

To be sure, the E-2C airborne early warning aircraft, with its compact cabin and crew of three radar operators tightly packed in the back of the fuselage, did not come close to the size envisioned by the maritime patrol community for their MMA, and it would necessitate a compromise in crew size, mission equipment, and range. But a Bethpage strategy was emerging based on the notion that the Navy might be persuaded to accept the compromise to obtain the cost savings that could be generated through commonality and economies of scale with the carrier-based air-craft. The CSA benefitted from the application of the same QFD process employed for MMA, but key mission performance parameters so differed that a yawning gap between MMA and CSA requirements soon forced the CSA out of Northrop Grumman's consideration for finding some alloyed blend of designs. By the spring of 1999 the Navy had come to a similar conclusion about merging mission requirements inside the CSA candidates and put the aircraft on indefinite hold.

After months of government analysis and informed exchange of ideas with the National Defense Industrial Association, the MMA initiative began a formal concept study phase with the release of the MMA broad agency announcement (BAA) in early 2000. The announcement was a solicitation to industry, intended to provide a contractual umbrella that permitted greater access to relevant government data and to deepen the open dialogue between industry and Navy officials that had already been established to a degree through the NDIA intermediary. At its most ele-mental level, the BAA sought to standardize metrics by which to gauge industry concepts. A set of canned tactical situations was appended to the announcement for industry simulation and analysis that broke out mission functions for broad operational descriptions of antisubmarine warfare—high and low, anti–surface warfare—high and low, surveil-lance/reconnaissance, and so on. Industry was expected to "play" MMA concepts in each tactical situation and report in their concept studies on the performance obtained. Navy evaluators would assess and select for continuance those that most closely conformed to the maritime patrol community's conception of a viable MMA candidate.

During the period leading up to the release, Frauenberger and Easterling had come in regular contact through the Navy and industry QFD sessions sponsored by the National Defense Industrial Association. They developed a mutual regard for each other's professional opinions. Since Northrop Grumman had no leading aircraft candidate in the mix, Easterling perceived Frauenberger as an unbiased voice of objectivity as the BAA was forming since he was not peddling a preferred concept in the manner of Boeing, Raytheon, and Lockheed. Similarly, Frauenberger was drawn to Easterling's openness to fresh thinking that occasionally departed from the conventional. These preconcept study dialogues were intended to provide the Navy not only better clarity on industry designs but also insights that could be used selectively to formulate the MMA mission need statement and to construct a draft requirements document. While it was understood from off-the-record comments from Navy participants that MMA would either be a remanufactured P-3 or a modified in-service airframe, within the NDIA forum a "clean-sheet" mentality prevailed. Not yet in Easterling's scan was a recently acquired Northrop Grumman product line offering performance that approximated some parts of the MMA set of mission attributes.

THE MISSION BEGS FOR TRANSFORMATIVE TECHNOLOGY

Frauenberger had become aware of the Global Hawk UAV's potential in broad terms, as had most of the advanced design staff in the several geographically dispersed division-level ATDCs. In the Northrop Grumman corporate office, the more forward-thinking staff members had given some thought to schooling the most innovative minds around the company in the unique capabilities of this new UAV product line. Through company-sponsored visits to the former Ryan Aeronautical in San Diego, Frauenberger became more deeply aware of the potential of the Global Hawk UAV. With the CSA initiative eliminated as a potential entry for Northrop Grumman into MMA contention, Frauenberger now wrestled with how best to inject an unmanned aircraft into the MMA mix—how even to raise the possibility of such a radical non sequitur for the Navy MMA team in its quest for a new manned aircraft. This would prove to be a fateful moment in the fortunes of BAMS and ultimately Triton for Northrop Grumman and for the Navy.

Although Frauenberger was making every effort to keep his supervisory seniors informed of his actions with the Navy, he was getting very little guidance at this point. His particular brand of irreverent individualism, confidence in his understanding of the Navy's requirement, and experience so far with the outcome of the modeling and simulation results from the canned MMA tactical situations emboldened him to make his move. At one of the NDIA-sponsored industry meetings hosted at Boeing, he looked for the opportunity to cordon off Commander Easterling for a frank discussion, feeling he could get a fair hearing since Easterling had been encouraging all industry participants to be creative in their MMA concepts. Frauenberger found his opening during a break in the proceedings and opened with a question: "Do you really want creativity, and would you be receptive to an 'out-of-the-box' broad agency announcement response from us?" When Easterling unhesitatingly answered yes, Frauenberger realized he had created a once-in-a-career opening, a potential franchise-expanding opportunity for Northrop Grumman by offering up a modified Global Hawk, a UAV system proposed for Navy inventory to augment the manned MMA for long-duration, wide-area, open-ocean surveillance missions.

The opportunity seemed real enough, and Frauenberger—pulse racing at the prospect of what could be just within reach—understood immediately that he had to convince his management and a bid board to move forward. What would become Northrop Grumman's approach remained a closely held secret at this point; Easterling and his Navy team would only become aware of the formal offering when response to the broad agency announcement was submitted to NAVAIR.

As the BAA for concept studies was readied for release in late 1999, it still contained no accommodation for, or even mention of, unmanned concepts. Immediately evident to Frauenberger on reading the BAA language was that a modified Global Hawk could meet only a subset of the performance requirements framed by the MMA tactical situations and that a proposal based on that narrow set of capabilities would be judged to be nonresponsive. With a copy of the announcement in hand, he returned to Bethpage to get in the queue for a formal bid board presentation, the primary purpose of which was to secure supervisory concurrence that the initiative conformed to the long-range business plan and came with enough bid and proposal monies to mount a credible effort at a response. The bid board was chaired by the individual heading the Bethpage division, a new transplant from the West Coast in a continuation of Northrop's practice of appointing loyal home legacy Northrop-nurtured leaders over Bethpage to reeducate the former Grumman workforce. To prepare for the bid board, Frauenberger convened a meeting with key experienced Bethpage advanced design engineers to weigh the options. The output of that meeting was an overwhelming recommendation to respond with the unmanned mission adjunct idea.

Consensus proved more elusive on whether the proposal would be for a Global Hawk or would more indirectly suggest Global Hawk as a surrogate for further analysis leading to an optimal high-altitude, long-endurance UAV solution. It was reasoned and agreed by all attendees that a modified Global Hawk, though not responsive on every score, was the best way to keep Northrop Grumman in the MMA game and, if well received, might be exploited to entice one of the other competitors into a teaming arrangement. Partnering with other MMA-interested companies remained an overriding concern at the senior management level of the Bethpage division and higher in Northrop Grumman, eager to ensure that a significant piece of airframe integration and assembly work—with higher profit margins and business base—could be extracted from MMA. At the bid board, $100,000 was sought as the minimum amount required, but the bid board chairman approved a response of only $50,000 for an anticipated contract of $500,000, implying that there was no obvious discernable future for Bethpage in going down this path. He cautioned that any response that proffered a Global Hawk for a Navy unmanned system risked contaminating the nascent Air Force Global Hawk program run out of the West Coast division of Northrop Grumman leading the UAV business portfolio. Not yet obvious to everyone at this point was the festering acrimony between Northrop Grumman's East Coast and West Coast divisions over "property rights."

With the bid board's acquiescence, Frauenberger was made the study lead, and Bethpage got to work. Long-time Bethpage ATDC hands were brought into a small proposal cell that included John Raha for configuration management, Jack Gaudet for simulation and analysis, Jim Tedesco and Laurel Longshore for life cycle cost, and two West Coast engineers closely associated with Global Hawk who lacked any discernable East Coast versus West Coast bias, Terry Barefoot and Bill Walker. A veteran Ryan holdover in Mitchell's business development organization was the key to their participation. Lapins and Frauenberger had, in those early cross-pollination visits, hit a resonant chord with Ryan's Norm Sakamoto regarding discussions about Navy variants. Sakamoto, grasping the potential and not yet fully sensitized to the interdivisional wrangling, footed the bill for Barefoot and Walker to join as system experts. The study team was placed under Maris Lapins, who was charged with developing a longer-term overall capture strategy that would guide the next steps once a concept exploration phase contract was won.

The BAA was worded in the most general way to encourage any concepts that were compliant with the MMA needs statement. At the initial acquisition milestone meeting in February 2000 to consider readiness for proposed concepts to move forward in the acquisition process, there was consensus among Pentagon officials that "the Broad Agency Announcement strongly indicated *only a manned land-based aircraft* would be considered acceptable. A compromise was achieved with the release of another Broad Agency Announcement, stating explicitly that the Navy would entertain solutions other than manned land-based aircraft."[4]

MMA proposals were turned in by early spring 2000, and Commander Easterling's program management office team realized that, in accordance with Pentagon dictates, special accommodation would need to be made for Northrop Grumman and any other respondents with radical one-off concepts. Easterling called Lapins and spoke plainly: while he was intrigued with the concept, he did not have adequate budget to cover the proposed course of study. Easterling saw a proposed study cost from Northrop Grumman of $390,000, interpreting that number to be for Bethpage's share only, which he further assumed was not adequate to cover the West Coast UAV group's contribution to the study. Lapins made clear that the $390,000 was a total study cost. Feeling out his developing yet untried relationship with Easterling, he asked very directly how much was available for the adjunct study. With Easterling's answer, an upwardly revised study was submitted. Easterling's deputy, Craig Nickol, took a second look at the original program name for MMA—broad area maritime and littoral armed intelligence surveillance and reconnaissance aircraft—and advised: "Let's drop the 'Armed' and change it to BAMS," which stuck. Northrop Grumman, the only respondent that did not propose a large manned aircraft, was asked to withdraw its original BAA proposal and resubmit in response to a second announcement, this time for BAMS.

The wording that reflected Easterling's intent in the second BAMS announcement was an unmistakable reaction to the prior conversations with Frauenberger. Although the MMA solicitation focused on manned aircraft, other "systems, or combinations of systems which meet the requirements detailed by the officially-approved Mission Need Statement" would be considered.[5] This was clearly meant to acknowledge the potential for an unmanned system as part of a larger MMA system of systems. Of primary importance was the elimination of any specific reference to antisubmarine warfare capability as a part of BAMS—only the vague assertion of undefined future potential. In June 2000 Northrop Grumman was the only awardee of a concept study contract under the BAMS broad agency announcement for $490,000 signifying the launch of BAMS in earnest.

Easterling, mindful of the prodding for transformation from layers of Pentagon bureaucracy, sensed immediately that here was his answer, a totem to ward off further intrusion and gain advocacy with senior defense officials for MMA. After more discussion, Easterling was sufficiently roused to grasp the implications of a future force structure mix that somehow included unmanned and manned aircraft for more economical operations. That Frauenberger—by then steeped in MMA operational requirements and Global Hawk capability—conjured a vision for unmanned adjunct support to MMA, and Easterling—until

Northrop Grumman is believed to be the only contractor studying the use of a UAV as part of the MMA solution. "We proposed that the Navy look at a hybrid system of manned and unmanned assets, where you would offload some of the missions from the manned to unmanned platforms, with the UAVs working cooperatively with the manned aircraft to provide a system of equal or greater effectiveness with less investment and less risk to personnel," explained Howard Frauenberger, BAMS study manager at Northrop Grumman.

—KENNETH SHERMAN, "Broad Area Maritime Surveillance Concept Advances," *Journal of Electronic Defense*, January 1, 2001

then resolutely focused on discovering an affordable manned aircraft concept—was able to comprehend Frauenberger's vision and readily embrace its potential proved the formative moment for BAMS and, ultimately, Triton. This, then, was the instant of conception, the point of consensus ad idem where minds met. It unleashed a wave of disruptive force that unsettled conventional, hidebound thinking in the maritime patrol community about its self-image and operational practices. It is not recorded in any official Navy documentation. As a historic milestone, it cannot begin to compare to that day in December 1903 when the Wright Flyer broke away from Earth; it was more akin to the moment when Wilbur Wright first conceived of wing warping as a means to control an aircraft's rolling motion about its longitudinal axis. The challenge for Easterling lay in carrying the idea forward in a limited time frame by crafting BAA language that awarded credit for unmanned adjuncts to the manned aircraft.

Under the MMA broad agency announcement, Boeing, Lockheed Martin, and Raytheon were all awarded concept study contracts at values ranging from $250,000 to $500,000 to be executed over a six-month period. Boeing was to examine the feasibility of adapting the 737-700, originally a civilian airliner that the Navy had adapted for use as the C-40 logistics freight hauler. Lockheed Martin would study an improved variant of the P-3C Orion, and Raytheon would explore remanufacturing the P-3C fleet. The aircraft that was to replace the P-3C was to be known as the search and attack variant, and the EP-3E Aries replacement would be the signals intelligence variant.

In July 2000 the Navy principals traveled to Bethpage to conduct the first contract kick-off for BAMS with Northrop Grumman. Commander Easterling opened with remarks on behalf of the Navy team, noting that they found the proposed manned/unmanned hybrid MMA/UAV concept "visionary," representing "revolutionary" thinking for the Navy. He expressed anticipation that his team had a lot to learn from Northrop Grumman through the final report that would emanate from the contract study, but he also was frank in saying that he wanted the contractor team to become as familiar with P-3/EP-3 operations in antisubmarine warfare, precision surveillance and targeting, and network-centric warfare as would be possible over the next several months of contract work. The Navy team also recommended that the study consider the antisubmarine warfare–high mission for the UAV. Exchange visits were planned that would permit the contractor team members to further immerse themselves in P-3 and EP-3 operations at site visits to Naval Air Stations Jacksonville, Florida, and Whidbey Island, Washington, while the Navy team encouraged the mid-term interim program review be held at Edwards Air Force Base or Palmdale, California, and try and have it coincide with scheduled Global Hawk test flights.

Easterling stressed that affordability was a paramount concern for the MMA program, and he challenged the BAMS contractor team to provide highly accurate and traceable cost estimates as part of the study output. From a Navy perspective, it was evident at the initial meeting and throughout subsequent reviews that the Northrop Grumman team intended to use the BAMS label to differentiate their product from MMA. This clearly stemmed from the Navy's awarding a study contract titled BAMS. Yet despite attempts by the Navy team to emphasize that there was no such thing as a BAMS program and that Northrop Grumman's efforts were part of the larger MMA program, Northrop Grumman persisted, reasoning that to subordinate its UAV concept to the whims of the maritime patrol community—especially when the depth of the culture problem could not yet be calibrated—was not prudent strategy. If there was an overriding concern for Easterling at this point, it was minimizing negative perceptions of the unmanned initiative in the wider maritime patrol community, and to do that by explaining the benefits and limitations of an HAE UAV adjunct to the MMA—with emphasis on the positive.

Easterling underscored several times during the session the need for a joint Navy and Northrop Grumman effort to socialize the hybrid concept by mounting an aggressive campaign to disseminate the facts of the concept to the community warfighters and to dispel disinformation and emotional bias as much as possible. His concern was driven at least partly by reaction to a comment from his chief engineer in the Navy program office, Joe Landfield, that he was encountering zero enthusiasm for the UAV throughout the maritime patrol program office, on the CNO's staff, and in the wider Navy hierarchy. Easterling shared Landfield's concern privately with Lapins and Frauenberger, adding his own emphasis to underscore the need for a socialization campaign that would cast BAMS in the best light as a contributing element of a networked maritime patrol force. If the BAMS principals had not yet contemplated this looming, palpable cultural resistance to the specter of a UAV entering the maritime patrol force structure, they were now becoming aware of it—even if their awareness was entirely unfounded conjecture. It was widely believed in the uniformed operator community that appropriations to sustain and modernize their manned aircraft operations constituted a zero-sum game with respect to new alternatives The collective view was that the UAV would draw down precious budget obligation authority intended for the MMA, compromising a successful acquisition of the community's manned aircraft replacement—not to mention the threat of a drawdown in aircrew billets as the unmanned aircraft inventory grew in proportion to the manned aircraft. By all reports, this was a highly emotional bias that held sway at all pay grades, from junior officer to commodore.

Here was the onset of a profound dichotomy: UAVs would likely never become a regular part of Navy inventory if they didn't enter as an adjunct, a mission augmentation or enabling subordinate partner to the crewed aircraft; budget for and acquisition of UAVs could simply not be justified in isolation with no connection to a mission area for which a manned aircraft was already the preeminent platform. Yet the prevailing warfighter culture militated against any introduction of UAVs that were viewed as threats to funding for manned aircraft and their crews. Easterling rightly stressed the urgent need for a socialization campaign to lessen warfighter anxiety and doubts. But it would have to be built on irrefutably persuasive logic expressed in a few key selling points to a highly resistant customer base.

As the concept study neared completion, a major Northrop Grumman reorganization orchestrated from the Las Colinas, Texas, headquarters greatly affected each of the four operating divisions of the reformed sector level organization, but none as profoundly as Bethpage. Apart from Bethpage, the concentration of UAV programs on the West Coast that had been grouped as an integrated program team now was recognized as the fourth division of the integrated systems and aerostructures sector. But in Bethpage, a pent-up wave of long-tenured retirements opened up key leadership positions. From a background largely defined by contracts management, Dave Stafford acceded to the position of vice president for business strategy and development. Philip Teel, now running Bethpage, entrusted Stafford with oversight and direction of the on-site and remotely sited business development staff, strategic planning for future investment and growth, and—perhaps most importantly—the former advanced technology development center and its technology forays into traditional and new markets. The realignment brought together in one organization under Stafford those individuals who would have the most to do with moving the BAMS concept forward: Lapins as head of the ATDC, now renamed the advanced systems directorate, developing future strategies for BAMS and MMA penetration; Frauenberger as the technical and programmatic sage for BAMS under Lapins; and Snowden, joined to Lapins for customer strategy and Washington, DC, marketing operations.

Lacking an extensive technical or operational background, Stafford's strength as a leader was his inherent gift for reading defense market trends and intuiting shifts in national security policy that, in turn, heralded emerging service budget and force structure priorities. After a prior stop as the business development lead for Bethpage's electronic warfare products and services, Stafford had found his niche. His overseers in the newly structured Northrop Grumman integrated systems and aerostructures sector leadership took notice of his quick success in cultivating think tank personalities and congressional staff members. He proved adept at building shaping campaigns that elevated the volume, if not the breadth, of advocacy for continued funding of legacy systems in the Bethpage portfolio. He perceived the BAMS concept study as an entry play for Bethpage to gain a steadier foothold in what could potentially be a much larger maritime surveillance market. Now, Stafford set forth a vision for that goal and how his organization would attain that market penetration by emulating his success in the electronic warfare franchise. Only by a comprehensive shaping strategy might entrenched positions be moved to accommodate a radically new approach. Perhaps his greatest influence on moving BAMS forward in this regard was his restraint, coupled with unwavering encouragement, in allowing Lapins, Frauenberger, and Snowden to form a tight collaborative team to create and execute the shaping campaign.

BAMS TAKES FORM

In the early drafts of the concept study report, several findings were taking form that bore promise: a notional HAE UAV could play a meaningful role in six of ten VP and VQ missions defined in the MMA tactical situations; potential existed to eliminate some forward P-3/EP-3 detachments by conducting long-range missions from the five identified HAE UAV deployment sites; including four unmanned aircraft in a composite P-3/UAV squadron could reduce the number of MMAs required from nine to six; and a Global Hawk derivative represented the most economical approach in cost and risk to fielding an adjunct UAV for BAMS with a potential initial operational capability in 2011.

The most head-turning leverage appeared in two findings. First, worldwide surveillance footprint coverage could be achieved by operating HAE UAVs from five of the eighteen P-3 bases then operational —a testament to the exceptional speed, range, and endurance of the Global Hawk airframe when paired with the optimum sensor payload for unprecedented surveillance staying power. The condition that would eventually need to be met for a high-flying UAV to supplant the manned aircraft in a surveillance mission was for 360-degree radar coverage and adequate electro-optic and infrared sensing for identification of surface targets from altitude. The second finding was that a hybrid force of search attack MMA variants combined with surveillance intelligence variant MMAs and HAE UAVs could provide equal or greater effectiveness than a force of 251 MMAs alone. The Navy program office set its inventory objective at 251 aircraft—an approximate one-for-one replacement of the P-3 and EP-3 fleets. Using Navy program office guidance of MMA cost of $100 million the reduced numbers of manned aircraft made possible by introduction of a dramatically less costly unmanned aircraft created an advantageous cost trade-off. Potential savings of more than $5 billion in total ownership cost over a thirty-year life could be realized by transitioning to a hybrid system maritime patrol force of manned and unmanned aircraft.

The final concept study reports from the four contractor teams were expected to inform an MMA analysis of alternatives being conducted by the Navy's preferred in-house study experts at the Center for Naval Analyses. Their analysis would, in turn, feed budget quality cost estimates into a Navy acquisition strategy that could withstand senior Pentagon scrutiny at an initial acquisition strategy milestone review. The MMA analysis of alternatives was "originally funded at $1.3 million and was to have lasted a year. In the event, it represented a $2.5 million effort and took almost two years. In addition to its original scope, the Pentagon's office of program analysis and evaluation insisted that more definition be provided on the role of space-based assets and UAVs (in further deference to the 'transformation mandate') in maritime intelligence, surveillance, and reconnaissance."[6]

The Center for Naval Analyses explored a range of concepts but in their final conclusion pointed to an MMA concept for a maritime ISR constellation of manned and unmanned airborne sensors that depended mightily on network-centric warfare, a vision gaining popularity in future warfare studies. A conclusion drawn from this by the Navy program office was that an acquisition strategy might best satisfy review authorities in the Navy secretariat and with the secretary of defense's staff that called for a "prime integrator" that would be responsible for integrating the various elements of the system. The condition that was set in the strategy was that the prime integrator need not be one of the airframe manufacturers but would need to demonstrate prime level competence. Northrop Grumman was concerned. This implied a directed subordinate role for BAMS and could easily strangle the HAE UAV concept when funds were needed for the manned MMA, particularly if one of the MMA competitors was selected also as the prime integrator. Here again was a cultural challenge that had to be tamped down to ensure the viability of the BAMS concept. The prime integrator role, however, would not gain the traction needed to become a reality, probably as much from objections lodged by the MMA primes as by Northrop Grumman.

As the study wrapped in late December 2000, Lapins and Frauenberger set time to meet with Snowden in Northrop Grumman's Rosslyn, Virginia, office to share the study findings and to begin mapping a socialization strategy. Snowden was openly an early doubter of the UAV adjunct idea, but he warmed instantly when presented with the executive summary of the concept study with several provocative assertions. The most salient points had been rendered in PowerPoint charts showing cost comparisons and world maps indicating UAV surveillance coverage. An underlying foundation of the concept study proved to be the results obtained by Frauenberger's leader for modeling and simulation, Harry George. He constructed scenarios that replicated current, real-world operations for P-3 squadrons that would ultimately resonate with the maritime patrol community, adapting and tailoring constructive simulations that compared, for example, surveillance results of fast-mover drug runners launching from ports in Colombia employing P-3s, then Global Hawks. What was revealed was markedly better performance for the Global Hawk. The results, when rendered in animated slides, were startling in their visual impact and would prove to be the decisive charts that would draw serious consideration by other doubters and fence-sitters in briefings to come.

Much as Easterling had, however, Snowden sensed that overcoming the ingrained community culture would be a steep climb. Together, for the better part of several days, Snowden, Lapins, and Frauenberger batted around a multipart contact strategy and timeline, with waypoints and desired outcomes at each of several intervals over a year-long execution timeline. Not having worked closely together before this concentrated give-and-take session (sometimes called whiteboarding) brought personalities and perspectives into sharper focus: they began to take the measure of each other and gauge each other's strengths and depth of experience for a collaboration that would need to continue—and deepen and strengthen—for months to come.

What emerged was an early consensus among the three on which responsibilities each would assume going forward and an agreement that Lapins and Snowden would take the socialization campaign on the road as the primary briefers. Snowden proposed an underlying construct of tying contracts and messaging to the requirements and programming/budget cycle known as the program objective memorandum (POM) build-up for the coming year, hitting POM stakeholders' lead time ahead of their issue development or interaction with the process. The central idea would be to build awareness and advocacy for the adjunct concept, hopefully obtaining positive, or at least neutral, inputs; to protect program funding requests originating from Easterling's program office team under the MMA heading; and, if possible, to create a standalone budget line item for BAMS around which energy and resources could be marshaled for institutionalizing it as a program of record. From Stafford's playbook, an outreach campaign would be added that pulled in consultants, think tanks, and, when ready, congressional staff. It would begin almost immediately with briefings to community operators and fleet commands charged with development of warfighting requirements shortfalls, working its way progressively inward toward the CNO's staff in the Pentagon and NAVAIR staff at Patuxent River, with eventual stops at congressional committees.

Before making the first appointment, Snowden declared the need for a trip to the "oracle," a consultant board composed of retired flag officers and retired senior Pentagon officials that Stafford assembled at least three times a year to scrutinize Bethpage marketing strategies. Their deliberations were to gauge the realism and comprehensiveness of campaign strategies and to recommend alternate approaches when deemed necessary. Several Northrop Grumman units and many defense companies ran similar consultant review boards to vet business development plans through some sort of sanity check. Bethpage's board of consultants was particularly attuned to BAMS and the strategy that was presented in that several of the consultants had been intimately connected with Global Hawk in their former government positions. Bethpage, through the effort of Stafford, had contracted with Technology Strategies and Alliances Corporation, the consulting entity founded and operated by retired Adm. David Jeremiah. As the vice chairman of the joint chiefs of staff, Jeremiah had supervised the advanced concept technology demonstration (ACTD) steering group. He had been closely involved in the source selection and funding for the Tier II+ ACTD, which yielded the Air Force Global Hawk. Jeremiah brought on to the Bethpage panel of consultants Larry Lynn, former director of the Defense Advanced Research Projects Agency (DARPA) during the ACTD implementation, and John Entzminger, the Tier II+/Global Hawk program manager at DARPA.

Six years earlier, these three had been the prime movers in steering the new ACTD process and in shepherding Global Hawk through its earliest trials in the Pentagon bureaucracy. They were joined on the Bethpage consultant review board by a seasoned, fleet-experienced group of retired flag officers that included Vice Adm. Jerry Tuttle, former deputy CNO for command, control, and communications, Vice Adm. Herb Brown, former Third Fleet commander, and Rear Adm. Riley Mixson, former deputy CNO for air warfare.

Snowden had already briefed this assembled group multiple times on legislative initiatives for appropriations plus-ups that would sweeten available funding for E-2 airborne early warning aircraft radar upgrades and EA-6B electronic attack aircraft modernization. This was an era when favorable congressional additions to the formal defense budget request could frequently be obtained with a convincing pitch to a congressman's personal staffers based on constituent interest or to professional committee staff seeking redress for Pentagon shortsightedness. In senior-level meetings chaired by Teel in Bethpage, Stafford tossed some favorable accolades Snowden's way, acknowledging his success in achieving these plus-ups, especially for fabrication and replacement of EA-6B outer wing panels and center wing sections. Overall, they amounted to at least $25 million a year for several years running, and they extended the vitality of the electronic attack force; Stafford also became aware of and reported on Snowden's original work to grow a radar demonstration lab on Makaha Ridge in Kauai, Hawaii, by tapping into Senator Daniel Inouye's largesse for $5 million to $10 million per year over a multiyear period. This had the planned longer-term effect of kick-starting a radar modernization program that would, after several twists and turns, lead to a much-revitalized advanced E-2 production program. This prior work had brought Snowden in closer coordination with Stafford, but it was Stafford's endorsement to the leadership that secured Snowden's reputation and permitted him an unfettered run at drafting and executing the BAMS legislative campaign.

In these days, the corporate head of operations for Northrop Grumman's Washington, DC, presence was Bob Helm, who had worked previously as the DOD comptroller and who was experienced with the inner workings of Congress. He would abide no contact with Hill members or staffers that he did not personally authorize or that was unaccompanied by his lobbyists. There were only a few individuals that he permitted to work with the lobbyists and, occasionally, even to work unaccompanied on behalf of their specific or governing sector or division. Stafford was one who enjoyed that favored status; by extension, Snowden was now accorded greater latitude in preparing and briefing Hill staffers.

But before exposing a BAMS socialization campaign to a thorough probing by outside consultants, it needed to be placed in the context of a larger strategy, which had to be vetted by Stafford and Teel. Their top cover would be essential to shield the BAMS campaign from unsolicited and unnecessary interference by the unmanned systems division, the sister Northrop Grumman division on the West Coast. The division may have been justifiably concerned about commitments Bethpage made to the Navy in its BAMS briefings. Their Global Hawk's future was not secured with the Air Force yet, and any promises made on the basis of an as-yet-unapproved production program for the Air Force would be premature.

Lapins drew the responsibility to brief a broad outline of a strategy to Stafford and Teel. He reassembled the core team of Frauenberger and Snowden to flesh out the elements of the larger strategy that could convince Stafford and Teel that there was an attractive and achievable business case that warranted their commitment to BAMS. The vision put forth was expansive but was deemed achievable: establish a growth path for the Bethpage division as the Navy's sole-source prime for serial production of marinized Global Hawks; exploit the PRB subsidiary's incumbent prime role in tactical support centers and mobile operations command center configuration design support to obtain a lodgment in the command, control, and communications for maritime patrol missions; and, with that vision, establish a goal for increasing Bethpage sales by several hundred million dollars at the end of the five-year strategic planning window.

The strategy to attain that vision was distilled to several interlocking objectives: leverage the maturity of the Air Force Global Hawk to bypass further Navy competition and move BAMS directly to a new system development and demonstration phase; leverage that success to a sole-source buy of several preproduction BAMS models; represent Bethpage to interested foreign military sales customers with maritime requirements as the preferred Navy prime for BAMS; involve Northrop Grumman's electronic systems sector in Linthicum, Maryland, in payload development to meet unique Navy requirements for over-water operations—especially for a more capable 360-degree field-of-view inverse synthetic aperture radar; and engage the Northrop Grumman mission systems sector (predominantly the former Logicon Corporation) in mission planning and training. Teel and Stafford were enthusiastic and endorsed the plan, with one addition from Teel. He demanded a stronger statement of purpose in exploiting the BAMS position to reenter the MMA contest. He envisioned a role as a major team member with another company for significant aerostructure assembly production and mission integration content.

At the time, the Navy program office was still considering an MMA/BAMS overarching prime contract that would have single-source responsibility for integrating payloads, connectivity, and operations for MMA and BAMS. It was clear to most in Bethpage with a modicum of industry experience that if such an approach were adopted, the MMA prime would rule, and BAMS could easily fall victim to inevitable MMA overruns. The capture core team would have to walk a very narrow tightrope in complying with the Navy's stated direction and, at the same time, not surrender autonomy and independence in determining BAMS' future. As is so often the case, the Navy's position would soften when the impracticalities of a single overarching prime became more evident. But for the moment, Lapins, Frauenberger, and Snowden had a green light on the strategy and top cover from their leadership to press ahead with the socialization contact plan.

STANDING BEFORE THE ORACLE

So it was in late winter 2001 that the consultant review board was first exposed to a multifaceted strategy to take BAMS beyond the concept study phase. Lapins and Snowden appeared as a tag team of briefers, with Lapins explaining the concept study report and strategy objectives and Snowden elaborating on implementation through the socialization and legislative campaign. As briefed, the socialization strategy was composed of messaging, audience, and timing inside the Navy; a parallel legislative strategy for institutionalizing a BAMS program element line item in the nearest congressional appropriation; and a plan for play in simulations, especially fleet battle experiments (FBEs), to demonstrate with hardware the warfighting utility that the maritime patrol community could expect from BAMS. The theory was that a socialization campaign composed of briefings alone would never turn heads so much as actual results from hardware put through its paces in highly visible and well-regarded FBEs. The consultants concurred with every outlined step and encouraged only more urgency in the execution. They were particularly anxious to see progress in exercises and simulations as the most likely means to sway a resistant maritime patrol community. Periodic updates would follow, but the primary briefing team of Lapins and Snowden soon began targeted contacts, occasionally joined by Bill Walker from the West Coast unmanned systems division at Rancho Bernardo who had recent Air Force U-2 experience and could best explain networked connectivity for intelligence products.

The first socialization presentation had to be to the newly founded Navy Warfare Development Command (NWDC), then in Newport, Rhode Island. According to then-CNO Adm. Jay Johnson, the command

The Global Hawk Maritime Demonstrator (GHMD), one of two RQ-4As acquired from the Air Force in 2005, is shown in flight over NAS Patuxent River, Maryland. Renamed BAMS-D, the aircraft lacks the prominent round bulge in the lower mid-fuselage that would accommodate a full 360-degree scanning radar, and the turreted nose sensor that would become a pronounced feature of the later MQ-4C Triton. *Northrop Grumman*

The most distinctive feature of the BAMS-D is its wingspan, shown here in planform perspective. At 131 feet tip to tip, it measures nearly twenty feet longer than a B-737 commercial airliner's wingspan. A single Rolls-Royce AE 3007 turbofan engine powers the BAMS-D to 60,000 feet altitude for 30+ hours. *Northrop Grumman*

A Northrop Grumman–owned Gulfstream aircraft flies past, the bulge underneath carrying the multifunction active sensor, or MFAS, for the BAMS proposal. The test bed aircraft with MFAS was a key element of a Head Start initiative undertaken by Northrop Grumman with company investment, intended to resolve risk items identified with their technical solution by characterizing reduced risk and cost avoidance for the Navy. *Northrop Grumman*

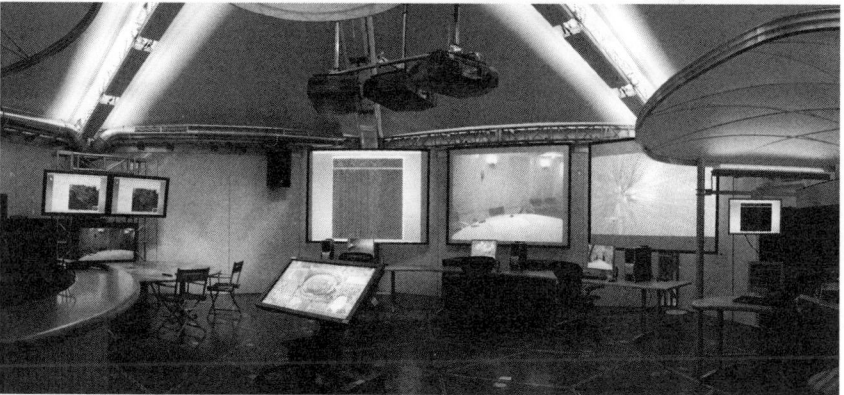

A second key element of the Head Start initiative was construction of a domed ground station populated with working displays and communications systems—seen here—designed to showcase the team's understanding of maritime ISR. In a scheduled live event in the so-called "podule," journalists from *Aviation Week & Space Technology* magazine were invited inside to witness transmission of radar data from the airborne Gulfstream test bed to the screens inside the podule. *Northrop Grumman*

Adm. Mark Ferguson, Vice Chief of Naval Operations, presides over the unveiling of the MQ-4C at Palmdale, California, June 14, 2012. Ferguson remarked, "and now the beginning of an unmanned tradition in our fleet with the unveiling of BAMS. History will record this introduction as a significant milestone in the second one hundred years of naval aviation." *Northrop Grumman*

At the unveiling ceremony in Palmdale, California, Navy flag officers with a vested interest in the successful progress of the MQ-4C program gather for a group photo. *From left*: Vice Adm. Al Myer, Chief of Naval Air Forces; Rear Adm. Bill Shannon, Program Executive Officer (Unmanned and Weapon Systems); Rear Adm. Bill Moran, Director Air Warfare (N98); Rear Adm. DeWolfe "Bullet" Miller, Director for Intelligence, Surveillance and Reconnaissance (N2N6F2). *Northrop Grumman*

With official remarks concluded at the Triton unveiling, guests and members of the official party gather in front of the aircraft on June 14, 2012. At this event the Navy announced the name change from BAMS to Triton. Two years later, the aircraft unveiled here would be the first to fly nonstop from Palmdale, California, to NAS Patuxent River in Maryland to begin system development and demonstration tests. *Northrop Grumman*

An MQ-4C Triton, the fifth production model, taxis at the Palmdale facility preparatory to its flight to NAS Patuxent River. For that cross-country flight, a mixed team of Navy and Northrop Grumman operators would execute takeoff from inside a ground control segment at Palmdale serving as the forward operating base, handing off control of the aircraft once established in flight to a Navy team inside a system integration lab at NAS Patuxent River. *Northrop Grumman*

Viewed in profile, Triton's most obvious differences from its predecessor, BAMS-D, are clearly visible: two ventral fins beneath the empennage for stability in yaw; the large, round-shaped pod on the underside that houses the 360-degree scanning active electronically scanned array radar; and the turreted electro-optic/infrared sensor turret on the chin of the aircraft. *Northrop Grumman*

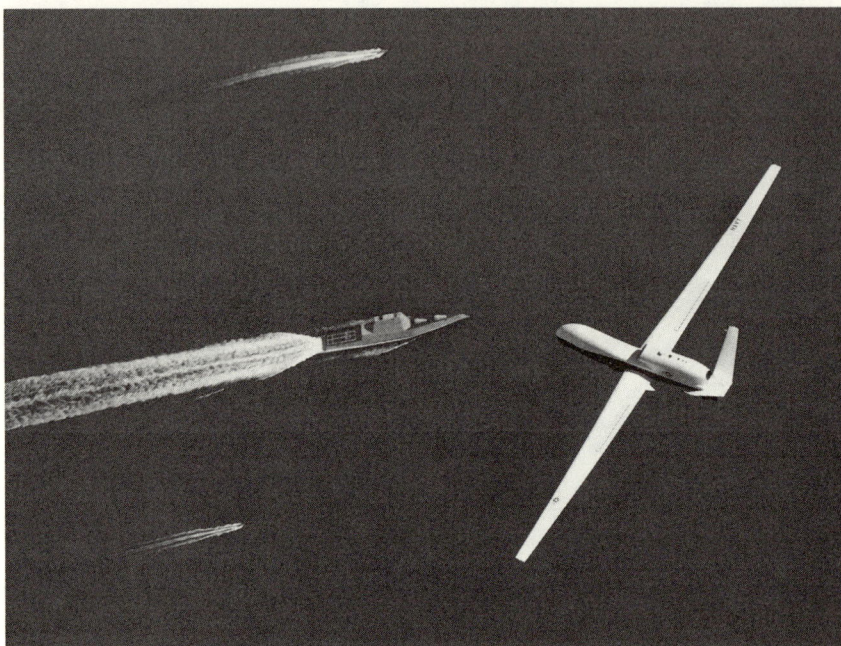

Triton overflies USS *Zumwalt* on the afternoon of the guided missile destroyer's commissioning in Baltimore on October 15, 2016. Three weeks before, the Triton program successfully passed its Milestone "C" acquisition gate, signifying its readiness to enter low rate initial production, coming fully eight years after Triton was determined the winner of the BAMS industry competition. Approval was also granted to re-prioritize program focus to provide for an early operational capability. *Northrop Grumman*

was intended "to work in concert with the numbered Fleet commanders and various labs and tactical centers of excellence for the development of concepts and the exploration of enabling technologies . . . meant to be the focal point for the development of Navy warfighting concepts . . . to be the center of innovation which will identify capability and acquisition requirements . . . including rapidly deployable networked sensors, and the use of unmanned vehicles in all domains."[7] NWDC did much of its work through the FBEs. As the Bethpage-developed BAMS socialization plan began in early 2001, planning at NWDC was already well under way for FBE-Juliet about eighteen months out in summer 2002.

Rear Adm. Robert Sprigg was the incumbent commander of NWDC, and Snowden made the appointment with his immediate staff to receive Lapins and Snowden in Sprigg's office to pitch a role for BAMS in the nearest possible FBE. When briefed, Sprigg was detached and indifferent. Despite the obvious congruence with FBE aims and BAMS capabilities, the opportunity apparently didn't merit his further consideration. It was not until several weeks later when Sprigg and Northrop Grumman sector head Ralph Crosby bumped into each other at a high school reunion and discussed the idea further that Sprigg experienced a change of heart. Snowden and Lapins were called back to NWDC to brief again, this time with Sprigg's extended maritime battle center staff. Near the conclusion of the brief, Sprigg became animated and demanded, "When can I have it?" His FBE-Juliet planners were taken with the possibilities and anxiously accommodated a Global Hawk appearance as a part of an experiment task to determine if an HAE UAV could be an effective source of information for ISR management. FBE-Juliet would be executed within a larger joint exercise, Millennium Challenge 02, but for the Navy component, an understanding of how a maritime operations center could handle dynamic tasking involving real-time intelligence/surveillance products from an input source such as Global Hawk was shaping up to be a critical question.

Once engaged with NWDC planners, this snapshot look into the future joint employment of Global Hawk and potential issues with exploiting its information revealed a leverage point—a key opportunity—that could be worked on behalf of BAMS. If a case could be made for additional but as-yet-unrequested funds to shore up this desperately needed capabil-ity—enabling Global Hawk information to be used effectively by Navy operations centers—then a congressional plus-up initiative might be in the offing. That struck a note with Snowden, but a few months remained to allow that thought to mature before being acted upon.

NWDC officers were not unfamiliar with Global Hawk's successes and setbacks with the Air Force. Beyond NWDC, what news of Global Hawk that had been digested by wider Navy audiences in advance of a BAMS briefing was usually interpreted to fit preconceptions about unmanned aircraft, their maturity as a new technology, their warfighting utility, or, more concerning for the socialization briefers, their suitability to augment or even supplant manned aircraft. Global Hawk's fortunes with the Air Force shadowed the Northrop Grumman BAMS briefers and distracted from their message as often as they reinforced it. Global Hawk had transitioned to its demonstration and evaluation phase, with the Air Force taking a more active role in exercise planning.

By spring 2001 with the Navy BAMS concept study complete and the socialization campaign just getting under way, Global Hawk was more than eighteen months into a planned twenty-one-exercise schedule and getting wider exposure with joint participants. Program sponsorship had passed to U.S. Joint Forces Command, which shielded Global Hawk to some degree from premature and arbitrary Air Force budget action. Navy Adm. Harold Gehman, head of Joint Forces Command, retained control over Global Hawk and ten other ACTDs, ensuring the newly hatched concepts got a full and fair hearing in major exercise play before entering into funding competition with service favorites. In April 2001 Air Force planners, in conjunction with Admiral Gehman's staff, were set to play Global Hawk in a major exercise known as Tandem Thrust, spon-sored by U.S. Pacific Command and designed to test the U.S. Seventh Fleet's ability to perform crisis contingency response tasks. In that year's Tandem Thrust, U.S. forces would exercise alongside Australian forces, with Global Hawk assigned to work with the opposing orange force. To preposition for its role in Tandem Thrust, Global Hawk needed to make its way to Australia.

CHAPTER 6

GAINING A FOOTHOLD

A large, complex sale doesn't consist of a single customer decision; it is a series of events. An event is a scheduled decision-making occurrence that influences the procurement outcome.

—MICHAEL O'GUIN[1]

The trip from Edwards Air Force Base in California to Adelaide, Australia, was accomplished in one nonstop flight, an unprecedented feat for an unmanned aircraft and one that would gain widespread notice for the UAV. In fact, in spanning the Pacific in a single flight, the UAV set a record as the first to do so and added luster to its growing reputation for long-endurance flight. But the long-distance flight in and of itself would not be the achievement that would open eyes and draw exclamations from exercise leaders. It took only days for the orange force, the opposition force in the exercise, using Global Hawk as an asset to grasp its capabilities to augment intelligence being gathered on the blue (friendly) force disposition. In one particularly revealing bit of intelligence, Global Hawk imaged the U.S. aircraft carrier USS *Kitty Hawk* acting as the command center for the blue force.[2]

In an instant, blue force complacency had been dashed. Their command center's location, which they believed to be concealed by its mobility far at sea, had been compromised. And importantly, orange force intelligence had been able to establish its location in a manner that provided precise geolocation suitable for targeting by orange forces. Seventh Fleet staffers were surprised and annoyed that their location had been so easily given away. The two BAMS briefers, however, were more than delighted to add that image to an introductory slide in their socialization briefings to Navy audiences. It immediately captured the undivided attention of those audiences for the BAMS study conclusions that followed. Perhaps just as important, the image of USS *Kitty Hawk* started circulating in the Pentagon, raising eyebrows among Navy leaders.

U.S. Joint Forces Command compiled grades of Global Hawk's performance in Tandem Thrust and other joint exercises in a military utility assessment that would be used to establish an official but preliminary set of desired UAV capabilities. Results were generally positive. With their release of that utility assessment, Joint Forces Command further recommended immediate transition to production, bypassing the formal engineering and manufacturing development phase established as a procedural standard for Pentagon acquisitions. This set off a round of contentious debate between Air Force and Pentagon leadership, the former insisting that the earlier ACTD phases had not provided a verified set of specifications that defined a production-ready aircraft and that, furthermore, no provision had been made for training, spares, and maintenance to permit fielding of the system. On top of those issues, the cost cap set for the original concept at $10 million unit flyaway price was now estimated to exceed $15 million and was still climbing.

Amid these developments, the operational side of the Air Force began to weigh in with their view of requirements, which had most to do with electrical power and open architecture to permit sensor upgrades. The beginnings of a B model Global Hawk that would follow the first Block 0 (ACTD) models and early Block 10 test aircraft were taking form but almost immediately began to run into headwinds with Air Force leadership. It became clear that the sensor package was proprietary and integral to the original configuration—meaning that other makers could not replace sensors and equipment models due to a very tight contractual arrangement between Ryan Aeronautical Company and Raytheon during the earlier ACTD vehicle design and development.

What was becoming more openly apparent was that even with the good news of Global Hawk's exercise performance, "both intelligence and operations branches of Air Force Headquarters remained opposed to the Global Hawk idea. The Joint Staff at the Pentagon continued to support the Global Hawk program, but unlike the Air Force, the Joint Staff didn't have to figure out how to pay for it."[3] To the extent that the inquietude spilled over to become more common knowledge throughout the Pentagon, it contributed to an aura of uncertainty about Global Hawk in the minds of the Navy staff. The Bethpage BAMS team was unde-terred—their singular focus at the moment was on socializing the broad concept with Navy influencers, operators, and decisionmakers. They could conveniently get by making only glancing reference to the closed proprietary sensor payload package, admitting privately that with Navy interest would come a demand for different sensors made necessary by maritime mission requirements. That issue would have to be dealt with in time but in the interim would remain an open wound between Bethpage and the unmanned systems (UMS) division in Rancho Bernardo.

On the socialization trail, the interaction with NWDC had become quite positive after the initial disappointing briefing with Rear Admiral Sprigg. At roughly the same time as the BAMS brief to NWDC, the maritime patrol program office engaged with the same experimentation planners and volunteered use of NAVAIR's testbed P-3 aircraft, known colloquially as the Hairy Buffalo, as an MMA surrogate to interoperate with the Global Hawk. NWDC was well pleased with the prospects and began drafting a set of experimentation objectives. BAMS, or at least a surrogate for BAMS in the form of an Air Force Global Hawk, was approved for play in FBE-Juliet. The BAMS team's socialization objective for obtaining early experimentation not only was met with FBE-Juliet but also was exceeded with NWDC's invitation to participate in its Global 2001 war game at the Naval War College that July. A member of the con-sultant review board with connections into NWDC ferreted out after the fact that "BAMS did very well, and that a consensus view had taken hold at NWDC that there was exceptional military utility to be gained with a long-endurance, high-altitude UAV—at least within the boundaries of the scenarios played." The first major milepost in a long chain of planned socialization events had been a qualified success. More mileposts were to come, but the BAMS team was encouraged.

By the time NWDC's Global 2001 war game wrapped, the socializa-tion campaign team had made multiple visits to brief the maritime patrol community at Whidbey Island, Washington; Kaneohe, Hawaii; and Jacksonville, Florida. Response was not uniformly negative but could not be in any way characterized as warmly supportive. This was the cultural "veil" that Easterling had been concerned about and that had prompted his request for a comprehensive socialization campaign in the first instance. More unbiased and enthusiastic response was encountered with the requirements staff officers at U.S. Pacific Fleet and U.S. Atlantic Fleet headquarters, and with Naval Air Forces Pacific and Atlantic Fleet staffs, all of whom took a more integrated view of collaborative and networked theater-level warfighting and could envision a BAMS contribution. For all of the positive resonance in the fleet briefings, the only stop that seemed to yield a less-than-friendly reaction was at Second Fleet headquarters, where Vice Adm. Mike Mullen presided over a briefing audience of more than a dozen captains and more junior flag officers. His response in the briefing was particularly negative, where every point made by the briefing team was met with eye-rolling disdain and cynicism. That said, it was a valuable result for sharpening the briefing style, creating more balance in the briefing content toward cost justification and incorporating additional modeling results and outcomes from Global Hawk exercise play.

ENTERING THE "RANCH," WHERE PROGRAMS LIVE OR DIE

As the campaign worked its way inside the Washington, DC, Beltway, Pentagon audiences were targeted and appeared much satisfied with the efficacy of the BAMS story, including naval and air warfare offices within OSD's directorate for defense research and engineering, the requirements directorate on the joint staff, and in those offices of the secretary of defense concerned with industrial capabilities. On the CNO's staff, stops were made at that office charged with integrating overall naval warfighting architectures and with effectiveness assessments of programs in scenario simulations. These assessments normally drove programming and resourcing issues to be taken before a flag-empaneled resources and requirements review board, culminating in decisions by a deputy CNO. Positive word of BAMS—at least in wargaming simulation results—was permeating Navy channels outside of the maritime patrol community, and it was generally well received.

By mid-spring, the stage was set to take the briefing before the director of air warfare on the CNO's staff, Rear Adm. Mike McCabe, whose own staff typically prepared the documentation for resourcing new program proposals. Two dozen air warfare staff officers, ranging from lieutenant to captain across the spectrum of platforms represented, assembled in the Ranch conference room to hear McCabe's reaction. Lapins briefed while Snowden acted as note-taker at the back of the room next to the section head for maritime patrol requirements, newly selected Capt. James C. Grunewald, USN. Grunewald barely tolerated attempts to spread the word on BAMS and was more than a little incensed that his boss, the two-star flag officer, consented to entertain this distraction from the BAMS contractor. Grunewald, ever supportive of his maritime patrol community, was grudgingly supportive of BAMS only to the extent that it made possible his MMA funding.

The argument had been made in closed staff briefings that a mixed force of manned and unmanned aircraft could be advanced to justify fewer new MMAs that otherwise would be viewed as unaffordable. This logic was identical to that put forth in the BAMS concept study, but Grunewald and others on the staff seized upon it as the smokescreen to pilfer MMA funds. Once MMA was well established, perhaps BAMS could receive more attention, but for now, the highest priority was MMA. Grunewald even produced a bar chart for his internal briefing that compared the relative inventory ownership costs of manned and unmanned adjunct systems, if for no other reason than to solidify MMA support. Unknown to or forgotten by Grunewald, Frauenberger had been provided a copy of the chart months earlier and had applied his characteristic level of rigorous analysis to update the cost numbers.

As Lapins moved through the first few briefing slides, McCabe remained impassive. If he had a preconceived idea about BAMS, it was not revealed by any outward expression until midway through the slides, when the bar chart went up. McCabe leaned forward in his seat, clearly focused on reading and understanding the numbers. He had seen a version of the chart in internal air warfare briefings, but here the cost divergence between an MMA inventory alone versus the mixed inventory was stark—and much more compelling. Grunewald leaned in Snowden's direction and said, in a voice that he meant to be heard throughout the conference room, "You guys are screwing me!" McCabe waved for Lapins to pause as he turned around to face his staff assembled in the conference room, asking, "Whose chart is this?" Lapins spoke, asserting that Northrop Grumman had developed the detail to conform to a template obtained from the air warfare staff.

Grunewald's reaction was the most outward manifestation yet encountered by the BAMS team of the deep-seated anxiety that gripped the maritime patrol community. It was cultural. The community had existed for a decade at the back of the priority stack for resources—primarily for tactical aircraft programs—while its relevance eroded and its inventory nosedived with fatigue issues and overuse. Now they assumed that BAMS was potentially fragmenting their new aircraft recapitalization plans to an even greater degree. Grunewald started to downplay the BAMS slide. McCabe, in a command that most in the room took as censure, told Grunewald to "be quiet. I know what you think, Grune; I want to hear what they have to say." This was a turning point for the collective perception of the CNO's staff about BAMS, and it gave license to McCabe's newly chartered UAV shop under Rear Admiral Kilcline to be more deliberate and proactive in their advocacy of BAMS. Emerging from that briefing, Lieutenant Commander Turner, Kilcline's lead staffer for BAMS, turned up the gain on his contact with Northrop Grumman to pull information he would need to fully develop a program proposal, essentially the official justification package that would be carried forward in the review cycle for new systems.

Snowden collaborated with Turner daily during most weeks throughout the late spring and summer, with less frequent meetings with Kilcline to shape a funding profile that ramped up to increasing numbers of BAMS systems: aircraft, mission control element vans, and launch and recovery system vans, which at this time both believed could be justified as a branch off the Air Force Global Hawk production line. Lapins turned to his team to fully develop the profile, which rested on entering a risk reduction phase in fiscal year (FY) 2002–03, followed by an abbreviated engineering and manufacturing development phase and then by low-rate initial production in FY2004 through 2007. These steps would logically proceed toward an initial operational capability in FY2008 wherein enough aircraft and ground control equipment, along with trained controllers and maintainers, would be on hand to permit a deployment that could sustain one maritime search orbit. Risk reduction would see the build and delivery of Navy prototypes that spiraled through payload updates, leading to fully deployable configurations for experimentation before a sole-source production decision.

Timing was such that a summer 2001 program proposal could not generate a funding appropriation before FY2003 some fifteen months out. To lead-turn that timing, Snowden conceived and went to work on a long-shot initiative to gain congressional plus-up funding in FY2002. The idea was to simultaneously generate congressional awareness while "institutionalizing" the BAMS funding mark as a measure of that congressional interest that could then be pointed to in subsequent Navy and Pentagon briefings. With approval from Stafford and encouragement from the consultants, Snowden then made the BAMS plus-up a part of his portfolio of requests. The head of Washington operations and the assembled lobbyists (one of whom typically worked Bethpage programs and would lead the way to Capitol Hill with Snowden in tow) further vetted these messages. Since this initiative would be presented outside the regular budget request and lacked any formal articulation from the Navy, the best hope was that the professional staff on one of the defense committees would assess this as evidence of inattention on the Navy's part. To Snowden's surprise, an advocate materialized on the House Appropriations Defense Subcommittee professional staff that agreed with the initiative and took action that materialized in the subcommittee report.

Broad-Area Maritime Surveillance (BAMS)—The Committee recommends $10,000,000 only for the development of a prototype Maritime Patrol and Reconnaissance (MPR) mission with advanced surveillance sensor equipment on a Global Hawk High Altitude Endurance (HAE) Unmanned Aerial Vehicle (UAV) for experimentation and for examining concepts of operation for use in conjunction with other MPR assets. The Committee directs the Navy to review and analyze results of previous maritime surveillance experiments with the Global Hawk HAE UAV prior to initiating BAMS. The Navy shall also submit a report no later than June 30, 2002 that includes its evaluation of the previous maritime surveillance experiments, the validity of the BAMS approach for MPR, an initial concept of operations for use of an HAE UAV as an adjunct to current MPR assets, and future year plans, including budget requirements, for continued development of the UAV approach to MPR.

—House Appropriations Subcommittee on Defense, Report 107-298, "Department of Defense Appropriations Bill for FY2002," November 19, 2001

While seemingly a small accomplishment, it came at an opportune moment a little more than six months into the socialization campaign when an outcome could not yet be measured, and it was received as an early vindication for the three BAMS disrupters—Snowden, Lapins, Frauenberger—and the rest of the Bethpage BAMS team. To capitalize on this foothold in the congressional appropriations process, Snowden would request for himself and Lapins an audience with Bill Stussie, the deputy assistant secretary of the Navy for air programs, to underscore the modest success with the plus-up and to add a sense of the congressional support behind it. The secondary purpose that Snowden and Lapins agreed to going in to the meeting was to tease out from the secretary how much of the work with the air warfare staff he had been made aware of and where he stood on going forward with the BAMS plan.

In that meeting, where only a military assistant sat in, Lapins and Snowden were relieved to find out that the secretary was very supportive and that he viewed the earlier DARPA-run ACTD competition as adequate for the Navy to branch off of for other service derivatives without further competition—much like the example of the Air Force adopting a Navy-developed A-7A Corsair II as its own A-7D without making the Vought company recompete for its own design. He noted that he was an advocate of Northrop Grumman–run competition inside the program for components and payloads to obtain optimum pricing and particularly sensors better suited to the Navy mission. He closed by noting that the plus-up would most likely be steered to a program element under the cognizance of the UAV program management office. That office in turn reported to the program executive for unmanned aircraft programs and weapons programs under Rear Adm. John Chenevey. That was a departure from the preexisting arrangement where BAMS had been nurtured inside the program management office for maritime patrol under the program executive for antisubmarine warfare.

The BAMS team departed, still unsure what to make of the last comment. The UAV program office, a repository of multiple classes of UAVs, was struggling without clear direction and lacked a mission mandate. The risk for Northrop Grumman and the Navy was that BAMS, lacking that clear mission and warfighting community linkage, would be allowed to wither. Without strong advocacy and lacking connection to the precepts of the BAMS concept study, the notion of its adjunct relationship to MMA would be lost. Yet if BAMS remained in the maritime patrol program office, it would forever be relegated to the lowest resource priority when stacked against the need for the manned aircraft replacement. Before the summer of 2001 was over, the maritime patrol program manager, Capt. George Hill, and UAV program manager, Capt. Lyn Whitmer, met with the deputy assistant secretary of the Navy for air to determine how best to integrate MMA and BAMS. Attending the meeting was Ron Rosenthal, a senior Navy civilian on assignment to the CNO's air warfare staff as the systems command expert on a detailed assignment at NAVAIR.

As the meeting opened, Rosenthal inquired as to the logic of BAMS residing in the UAV program office. His larger rhetorical question, and one that went straight to the heart of the more strategic issue for the Navy, was this: as UAVs began to make inroads into all mission areas, would the Navy's intention be to assign all UAV programs to that office? Or would it make more sense to recognize UAVs as a technology rather than a capability and assign the technology on a mission-centric basis to those manned aircraft program offices that were the acknowledged mission experts? The logic was irrefutable, but the secretary was nonetheless more interested in leaving UAVs grouped in the UAV program office as a management convenience.[4] Thus, from this moment, BAMS and MMA became increasingly detached from one another, and the vision of BAMS adjunct operations inside an MMA-populated maritime patrol community dimmed for the next several years. Still, the news that the Bethpage team was receiving from the UAV program office was upbeat and revealed a growing level of confidence and satisfaction with BAMS that had to have been coming from private Navy discussions.

In only their second meeting with the UAV program manager in January 2002, Hill and Whitmer described to the Bethpage team a two-phase BAMS program plan—a demonstration phase and a production phase—based on a sole-source award to Northrop Grumman. The production phase was to be a POM 2004 start, with a first-quarter FY2004 award.[5] This was a significant development; to bridge the acquisition process and take BAMS directly from a concept study to a sole-source production program would have been an unprecedented coup by BAMS disrupters Snowden, Lapins, and Frauenberger. Alas, it would prove to be more elusive than described in that meeting.

As the Northrop Grumman BAMS team worked intently with Rear Admiral Kilcline's staff throughout the summer and early fall of 2001 to construct a defendable funding profile for the programming and budgeting review cycle, the MMA team worked in parallel to secure funding for an initial acquisition milestone review and to sustain their momentum for manned aircraft recapitalization. As their MMA profile solidified with greater fidelity, their nonrecurring bill for MMA was approaching $3 billion. It was increasingly apparent that the director of air warfare, Rear Admiral McCabe, had no intention of funding the signals intelligence variant of MMA and, further, that funding for BAMS was becoming problematic.

When the Navy rendered its budget request to OSD, different priorities and viewpoints in that office governed the relative importance attached to UAVs under the mantra of transformation. In the budgeting process devised for ascribing secretary of defense preferences over service budgets, program budget decision number 732 was issued that established BAMS as a program of record with reprogrammed FY2003 funding of $383 million. This funding partly bankrolled the purchase of two Global Hawk platforms from the Air Force production line with additional sensor suites optimized for maritime surveillance, along with associated command and control systems. These systems were to be delivered in FY2005 in order for them to participate in FBE-Mike. To disassociate these aircraft systems from BAMS, which had by now become too closely identified with Northrop Grumman, a more generic title of Global Hawk Maritime Demonstrator (GHMD) was adopted. It fell to the Navy UAV program office to procure GHMD systems using existing Air Force Global Hawk low-rate initial production systems and engineering development phase vehicles.[6] The performance of the GHMD systems would be expected to inform any preproduction configuration changes required to prepare for the ultimate serial production program, which at this time was still contemplated as a directed sole-source procurement.

Less than two months after the program budget decision was issued by the secretary of defense, MMA moved forward to another senior

Pentagon review to assess readiness for exiting the concept study phase and entering the next acquisition phase. The Navy's position, presented at the review after rigorous analysis throughout the fall and into the winter of 2001, was that the EP-3E Aries signals intelligence replacement was not affordable as a part of MMA. There was not a great uproar, as most of the principals at the senior Pentagon review level saw no urgency in linking the replacement of the P-3C with the signals intelligence platform. With Rear Admiral Kilcline, McCabe's successor as director of aviation plans, and Rear Adm. Richard E. Brooks, the commander of maritime patrol and reconnaissance forces, in attendance, these

> Air Warfare representatives made the point that while the Navy was committed to full funding of MMA, the UAV piece [BAMS] would remain above core [*un-funded*, though GHMD was carried forward as a funded demonstration effort preceding BAMS] given its lack of definition. MMA could go forward while making provisions for UAV control from aircrew inside MMA; definition of the UAV was not required for definition of MMA. Nonetheless, it was agreed that while MMA would have its own operational requirements document and the UAV its own document, by the next acquisition milestone review there would be a capstone requirements document that tied the platforms together.[7]

MANNED AIRCRAFT HOLD SWAY

In early April 2002 Snowden was summoned to the Pentagon for a private meeting with the deputy assistant secretary of the Navy for air on the topic of GHMD and the sole-source production program that was thought to follow. What the secretary revealed in that discussion was that there would be no sole-source procurement to follow GHMD, that other interested defense firms had asserted to him that they possessed equal ability to produce such systems, and that the Navy had to declare an open competition for BAMS or risk formal protests. Not revealed by the secretary were the identities of the defense companies that approached him, but Northrop Grumman's inside-the-Beltway intelligence sources and consultants pointed to Lockheed Martin and General Atomics as the "aggrieved" parties. Northrop's position, reiterated by Snowden with the secretary, was that competition had already occurred with the DARPA Tier II+ effort; other companies had taken their best shot, and Northrop Grumman was judged fairly and accurately to have the better system. But the secretary buckled to the pressure of looming protest controversy and directed that all references to sole-source procurement be silenced.

GHMD experimentation would continue, but results would be available to all contractors on an equal footing to inform their own BAMS analyses. Officials in the UAV program office mouthed reassuring words that BAMS and MMA were still linked, but in fact, "BAMS had become an orphan without an established warfighting community of interest. While the BAMS operational requirements document (once issued) mentioned an adjunct relationship to MMA, it made no effort to define the relationship or articulate an adjunct requirement; rather it defined a system independent of MMA."[8] For the moment, all due effort at Bethpage was focused on preserving the GHMD for which funding appeared likely, and deferring, stalling, or killing a follow-on BAMS competition. The Northrop Grumman consensus view—Bethpage and Rancho Bernardo divisions together—was that no competitor platform and associated ground control systems in the industry were anywhere close to the maturity of the Global Hawk system and that the Navy was terribly misguided in needlessly adding several hundred million dollars to the program to bring competitors into a competitive range.

Even so, in early 2002 the BAMS team offered the UAV program office and the maritime patrol program office an opportunity to jointly take in and digest the output of Bethpage's noncontract technical investments in BAMS system development throughout 2001 through a series of technical interchange meetings. In January the work on concepts of operation and payloads was presented, followed the next month by presentations on air vehicle issues and the ground control segment, and wrapping up with modeling and simulation in April. These meetings exposed a number of unsettled questions bearing directly on the configuration of the GHMD system that urgently needed resolution for contracting to go forward. Their unsettled nature also presented an opening for Bethpage to steer the discussion to a more suitable long-term solution for Northrop Grumman and, as would be proved, for the Navy as well.

One critical issue was the selection of the radar for GHMD. It was readily apparent that the Air Force's Global Hawk radar was ill-suited to Navy needs for wide area ocean surveillance and ship detection and classification from altitude. Navy audiences were enamored of a radar system known by the designation APY-6, developed by Northrop Grumman's electronic systems sector. Yet the official Navy position on radar selection was to make it subject to open and fair consideration of all possible radar configurations in GHMD. The APY-6 radar had a long gestation in Navy labs and exercises that traced its origin and baseline technology to an earlier radar planned for the short-lived Bethpage-designed A-6F carrier-based attack aircraft. That radar then evolved to an improved design that Navy testers mounted in a pod and slung under the S-3 aircraft, calling it by the nickname Gray Wolf.

In its final state of evolution, the radar again morphed into the APY-6, finally wrung out in fleet experimentation when mounted in the special P-3 test aircraft nicknamed Hairy Buffalo. This long legacy of testing had yielded a system that now featured a moving target indicator, synthetic aperture to create two-dimensional imaging for finer resolution, and inverse synthetic aperture to exploit target movement to enable ship classification, all while maintaining the ground-mapping, all-weather capabilities of its legacy forebear radars. It now provided simultaneous detection, target acquisition, and low-level terrain information, and it could detect moving targets in clutter. The APY-6 had a synthetic aperture resolution twenty times higher than that of its predecessor configuration and a moving target indication ten times better.[9] The Office of Naval Research had already paid the electronic systems sector under contract to develop an accompanying electronically steered array antenna for the APY-6,[10] and this proved to be a most compelling feature for an already highly capable radar system—compelling because mounted on a Navy GHMD, the ability to scan electronically in a 360-degree uninterrupted sweep was much more attuned to Navy operations over broad areas of ocean than the side-looking, limited-field-of-view mechanical sweep that the Air Force Global Hawk offered. The electronic scan eliminated moving mechanical parts required to reposition the scan of the Air Force Global Hawk radar, and the 360-degree sweep opened up searches in all directions on command over open ocean.

Herein was the root of a seemingly intractable problem for Northrop Grumman. Tracing back to the origin of Global Hawk under the DARPA ACTD, Ryan Aeronautical Company had made a pact with Raytheon for a payload suite that integrated wholly into the Global Hawk flight control software. In effect, the Raytheon radar was inseparable from the vehicle. This was an anachronism in the new age of open architecture and would likely not have been permitted under DOD's latest acquisition instruction implemented in October 2000. The UMS division in Rancho Bernardo was concerned that the Bethpage division was encouraging the Navy to break out the radar and other payload systems, as the disruption could have a destabilizing effect on the Air Force Global Hawk program. This would further roil relations internally between the two divisions, but the issue would ultimately be settled by the Navy. A radar risk panel was convened with membership from Northrop Grumman and UAV program office to determine a way forward. The panel recommended deferring APY-6 installation to the second GHMD aircraft so as to not put a 2005 fly date at risk. However, the issue of the choice of radar for BAMS would continue to boil up as a hotly contested debate for another year.

In an effort to bring some order to UAV requirements going forward, the director of aviation plans staff under Rear Admiral Kilcline directed the drafting of an overarching roadmap. Beyond transmitting Navy intentions over the five-year defense planning window for the family of UAVs, its primary purpose was to make clear its commitment to buying two Global Hawks to assess their utility in a maritime environment. It officially applied the moniker Global Hawk Maritime Demonstrator to the two aircraft now funded and then described a BAMS program as a major competitive UAV program that was expected to go directly to a second acquisition milestone review in the second quarter of 2004. Described in the most recent DOD instruction, the milestone review would take BAMS from a presystems acquisition condition directly into systems development and demonstration, an acquisition phase roughly akin to engineering and manufacturing development in the previous regulation: "Independent of

To support its development of the BAMS concepts of operations, the Navy has proposed the acquisition of two Air Force production Global Hawk UAVs and a ground station for nearly $150 million in its FY03 budget. "That will allow us to get two experimentation vehicles out there to determine what this Broad Area Maritime Surveillance UAVs role will be," Captain Rand LeBouvier (identified as aviation systems branch in OPNAV) said.

—MARC STRAUSS, "Navy Outlines Future UAV Strategy," *Defense Daily*, February 20, 2002

MMA, the absence of formal stand-alone concept study and advanced development activity would continue to surround BAMS in ambiguity. Questions about technical performance, redundancy of mission function with MMA, communications infrastructure, and especially the placement of BAMS in a larger architecture of networked Intelligence, Surveillance and Reconnaissance systems were yet to be answered."[11]

Speaking at an annual convention of the Association for Unmanned Vehicles Systems International in Washington, the CNO's staff representative from air warfare, Capt. Rand LeBouvier, described for media representatives the UAV strategy just approved by Navy Secretary Gordon England: "We're not buying UAVs for UAVs' sake. . . . We are buying capability."[12]

His statement was made in contradiction to earlier rhetorical questions asked by many Navy officials, including Ron Rosenthal, the civilian detailee in air warfare from NAVAIR, who questioned whether BAMS was a capability or truly a technology that had application to potentially several mission areas as an adjunct to manned aircraft operations. The ideological premise of the BAMS concept study by Northrop Grumman had now been turned on its head, presumably as a pretext to restart the clock on BAMS in view of the lack of priority and funding to sustain any momentum. The BAMS team remained undeterred and continued on the preplanned and approved socialization roadmap, even as a sole-source production program (beyond GHMD) seemed to be slipping away. If its utility could be proved, GHMD could be an unrivaled source of decisive win themes in a coming BAMS competition, but a progression of yet-to-be-written documents defining the Navy's preferred BAMS attributes would need to be shaped toward GHMD features.

That effort was already under way by Lapins and his team in their GHMD preliminaries with the UAV program office wherein the team delivered a "Performance Specification defining desired radar threshold and objective requirements. This specification would be very much in evidence in later BAMS documents."[13] By the time the Navy released its solicitation to industry for BAMS in the fall of 2002, the BAMS teams had most of the documentation completed for its response, presenting a detailed description of the BAMS system concept and highlighting for the Navy the most challenging system elements, one of which was the importance of having a Track-While-Scan/Classify-While-Scan radar mode. The need for this feature was verified in a set of more stressing tactical situations analyzed by the Bethpage team that went well beyond customer expectations. Lapins and his team noted that the Navy responded by "calling these features out specifically in their performance-based specification."[14]

Lapins and his team noted also that the UAV program office appeared deeply invested in funding a tactical control system—the set of data links and consoles for controlling the airborne system—for a family of UAVs that would likely include BAMS. A tactical control system envisioned by the Navy program office would be woefully inadequate to the needs of BAMS and, from a purely self-interested Northrop Grumman business perspective, would allow another company to gain a significant foothold in the BAMS control architecture—possibly developing their own competitive win themes around their own proprietary tactical control system or, worse, involving themselves in rewriting software code for systems actually installed on the BAMS aircraft for the coming competition.

The team leaned into the issue and began a "reeducation" campaign with the UAV program office principals, the most satisfying outcome being that the tactical control system was not mentioned in the BAMS specification. What did get meaningful mention were elements such as the joint mission planning system and the global command and control system–maritime, which were integral to the BAMS mission control system architecture defined by the Northrop Grumman team. Finally, when the Navy went to a government-funded research center, MITRE Corporation, later that year to lead most aspects of the official analysis of alternatives for BAMS, the team focused again on providing a compelling presentation to MITRE on the methodology employed for evaluating BAMS sensors, including due regard for both air vehicle and mission payload performance. Bethpage requested an important second appearance with MITRE that allowed the team to walk through the costing of its methodology and the output; "The objective in both briefings was to attempt to introduce the notion of 'best value,' and evidence suggests it had the desired effect."[15] Best value would be adopted by the UAV program office at key junctures in the program through the major competitions to come.

Through the late spring and into the summer of 2002 Snowden continued his weekly activity on the team as the campaign leader for developing the briefing and the suggested language for congressional marks, those add-ons that would hopefully sustain—and increase—the funding appropriations from the previous year. Accompanied by the corporate lobbyist who had been assigned to work Bethpage legislative issues, Snowden made the rounds among the personal and professional committee staffs of the four defense committees prior to their marking period.

The interests of personal staff members and professional committee staff members generally diverged along very predictable lines, requiring tailored messaging, talking points, and briefing slides for each group. Whereas the personal staff was primarily focused on constituent concerns in their members' home districts, such as defense program jobs and infrastructure, the committee staff was more often focused on vetting programs against stated requirements, assessing funding accuracy and realism, and discerning troubling trends in program execution that could point to under- or over-spending of congressional appropriations. The committee staff normally hosted a visit from the military service representatives prior to the industry contractors getting their appointments, and it was not uncommon for the committee staff to use those visits to play one against the other to uncover a line of interrogation to be used in hearings. Not surprisingly, the central point of many congressional briefings conducted by Snowden was the value of BAMS as an enabler for the MMA. His talking points were organized to underscore for committee staffers the importance of robust funding for BAMS to fully develop its potential to offset numbers of MMAs, thereby saving the government the funding that would be required for one-for-one replacement of more costly manned maritime patrol aircraft.

A principal objective of the legislative campaign this time around was to build on the awareness established the prior year and particularly to create the sense that the Navy needed a little encouragement to act with more alacrity with regard to BAMS. The more the program was delayed by Navy indecision and lack of priority, the more time that was conceded to competitors to gain parity in the coming competition. Interest was again expressed from the House Appropriations Subcommittee on Defense staff—especially in the person of professional staff member Betsy Philips—that yielded a mark-up in the committee's FY2003 defense appropriations bill language.

Competitors, however, now had their own lobbying effort in gear, and developments over the preceding year had given them much inducement to influence congressional marks. In particular, the industry prime contractor for the tactical control system was well represented in

---

BROAD AREA MARITIME SURVEILLANCE (BAMS) UNMANNED AERIAL VEHICLE (UAV)—Despite its obvious support of the Navy's planned BAMS concept exploration and experimentation, the Committee is concerned about the lack of specificity and documentation provided thus far by the Navy. It appears the Navy continues to refine its plans for Phase I concept exploration, which is the basis for the fiscal year 2003 budget request, as well as the Phase II follow on development and testing program. Therefore, the Committee directs that the Navy submit, by February 1, 2003, a detailed report on the BAMS UAV program. At a minimum, the Committee directs that the report address the total program objective, the operational requirements documentation, the spiral development and testing milestone plan, the applicability of Department of Defense acquisition regulations, the training and support requirements, and the basing strategies for the system. The Committee also notes that the Navy has indicated that despite its plan to spend $24,000,000 on two air vehicles that would be delivered in 2005, the Air Force has not made a firm commitment to that delivery schedule. The Committee directs the Air Force to ensure that the air vehicles and other support equipment necessary for the Navy to proceed with the BAMS UAV program, is provided in accordance with the current schedule.

—House Appropriations Subcommittee on Defense,
Report 107–532, "FY2003 Defense Appropriations Bill," June 25, 2002

UNMANNED AERIAL VEHICLE TACTICAL CONTROL SYSTEM—The Navy is purchasing Global Hawk UAVs in fiscal year 2003 with the ultimate goal of integrating them into the TCS. The proposed fiscal year budget for TCS, however, does not fund such integration. Therefore, instead of using TCS to support the Global Hawks, the Navy now plans on using the existing, dedicated Global Hawk ground stations which are designed to work exclusively with Global Hawks. The committee believes that integration of Global Hawk into the TCS should occur as soon as possible to ensure TCS commonality within the set of Navy UAVs and recommends that $10.0 million be added to PE 35204N for this purpose. Furthermore, the committee urges the Air Force to work with the Navy to support the Navy's TCS activities.

—Senate Armed Services Committee, Report 107-151,
"FY2003 Budget," May 9, 2002

Congress, and the report language suggested that they were beginning to have some effect.

THE ROYAL AUSTRALIAN AIR FORCE WEIGHS IN

Foreign interest in the BAMS concept was beginning to stir in mid-2002, notably in Australia, and would act as a spur to U.S. Navy officials to remain focused on a sensor payload more in line with long-dwell maritime needs. It also would lay bare—again—simmering tensions between Northrop Grumman's Bethpage and Rancho Bernardo divisions. The Royal Australian Air Force (RAAF), a cooperator of the P-3 with the U.S. Navy, had taken a keen interest in MMA as the logical follow-on to its own P-3 inventory. An RAAF officer was assigned to the U.S. Navy maritime patrol program office as a liaison for the Australian Ministry of Defense (MoD). Through that officer, the RAAF was apprised of the BAMS concept study, and the Australian MoD was now interested in undertaking a similar study tailored to their maritime needs. Northrop Grumman's UMS division in Rancho Bernardo had already launched a marketing campaign in Australia for a possible foreign sale of the Air Force Global Hawk to the RAAF.

In short order, the MoD commissioned a contingent of eight RAAF officers from their maritime patrol group in Adelaide, Australia, to visit the United States on a fact-finding investigation. They were to visit both the East and West Coast divisions of Northrop Grumman's aircraft sector because they correctly perceived that the U.S. Air Force Global Hawk locus was Rancho Bernardo and the U.S. Navy study had been conducted out of Bethpage. Lapins and Frauenberger hosted the MoD delegation in Bethpage and provided a concept study overview, with rationale, and results that pointed to a more capable lower cost maritime surveillance force than mixed MMAs and UAVs. A week later, the MoD delegation moved to Rancho Bernardo and, upon learning that Bethpage representatives had not been invited to participate, insisted that Lapins and Frauenberger attend before the meeting could commence. The senior MoD principal made a point of sharing their chartering document with Frauenberger once he arrived; however, no other Northrop Grumman people were given access, signaling a note of trust already conferred on Frauenberger and the Bethpage site that may have been lacking with the West Coast group.

Within several weeks, a formal request arrived from the MoD for a quick-response BAMS-like concept study. At this point, the only contract mechanism for such a study was an existing contract between the MoD and the U.S. Air Force program office with Rancho Bernardo, yet the MoD insisted that Bethpage people be involved in a significant way. While waiting for contract negotiations to finish, the MoD invited several Northrop Grumman people to visit Canberra for a multiday review of a half-dozen classified Australian tactical situations. Again, the MoD felt compelled to insert itself to ensure that Frauenberger was invited to join the visit. Ultimately a statement of work was agreed upon, with a 60 percent/40 percent share split between UMS at Rancho Bernardo and Bethpage. Over approximately six months, the MoD was prepared to spend $2 million for what was, in effect, a second BAMS concept study. Study execution was remarkably smooth, despite the repeated insistence from the Rancho Bernardo participants that the MoD use the side-looking radar then installed in Air Force Global Hawks. The MoD was adamant that they would use only a 360-degree capability if they elected to go forward with a Global Hawk–based system. The MoD study wrapped up in August 2003 with a two-day outbrief in Adelaide. The overriding conclusion: MoD needs could be best met with twelve BAMS vehicles based in Darwin but commanded and controlled from Adelaide.

Australian MoD interest in BAMS never waned appreciably from that point. However, for the maturity they required in the system, they looked to the U.S. Navy for assurances before committing serious investment. The progress made by the small Bethpage BAMS team with the Navy concept showed spectacular results in the first two years, but now the inertia of the Pentagon acquisition system, the rising specter of industry competition, and the culture issues nursed by the maritime patrol community were retarding BAMS forward motion. Nonetheless, progress continued throughout 2003, driven by the persistence of the Bethpage disrupters, albeit not on the timetable envisioned by the BAMS team with socialization campaign results in 2001 and 2002.

In February 2003 Northrop Grumman was awarded a $185.2 million contract to initiate GHMD to be equipped with the sensor suite carried in the Air Force Global Hawk. The minor addition was the inclusion of software modes for inverted synthetic aperture radar capability and of a highly capable avionics set that contained radar warning receiver and electronic signals characterization for electronic support measures in the form of the LR-100. An APY-6–like radar, signals intelligence capability, and turreted electro-optic and infrared ball were deferred to the later BAMS configuration, on whichever vehicle that competition yielded.[16] BAMS documentation released by the Navy nearly simultaneously with the GHMD contract award to Northrop Grumman gave competitors the performance metrics to which they needed to design and the opening they sought to craft discriminators for the coming competition. It called for an operational range of 1,500 nautical miles with twelve hours on station time and a payload suite comprised of radar, a communications relay, a signals intelligence package, and an electro-optics and infrared package. By spring General Atomics was marketing hard its Predator B extended-range variant in anticipation of an industry solicitation later in the year, following completion of an operational requirements document then under review by the fleet.[17]

BAMS competition aside, GHMD was moving forward. In February 2003 the Navy awarded Northrop Grumman a contract for modifying and delivering "two Global Hawk air vehicles, two Integrated Sensor Suites (ISSs), two Launch and Recovery Elements (LREs) and one Mission

Control Element (MCE) for a planned 2005 maritime demonstration by the Navy."[18] The contract was awarded and managed by Air Force Aeronautical Systems Command—analogous to the Navy's NAVAIR—at Wright-Patterson Air Force Base on behalf of the Navy and specified integration of maritime sensors into two Block 10 Global Hawk air vehicles. Terms of the contract specified that the GHMDs were to demonstrate high-altitude, long-endurance surveillance capability that met the Navy's critical requirements. Since the contract vehicle relied on an existing relationship between the Air Force Global Hawk program and Northrop Grumman's UMS division in Rancho Bernardo, the UMS division logically became the managers of the GHMD effort. Bethpage didn't resist that determination and instead turned its primary focus to the coming BAMS competition.

The release of the BAMS operational requirements document in May 2003 refined and codified previous efforts by the Navy at defining BAMS performance attributes. Lapins and his design team prepared a self-evaluation against the document for presentation to the sector leadership by the next month, indicating gaps and compliance. Lapins presented a briefing that assessed his team's BAMS system concept to be "green" with respect to all key performance parameters and most threshold requirements—by even cursory analysis, the Northrop Grumman offering would meet or exceed most objective requirements. The evidence was plain, however, from two years of BAMS tactical situation analysis that the threshold requirements for air vehicle performance had been set sufficiently low to permit competitors such as the Predator B extended-range concept to be acceptable and thereby guarantee competition. By September, with threshold and objective requirements firming up, General Atomics and Lockheed Martin teamed to go after BAMS, with Lockheed Martin stepping up as the prime integrator to handle the system architecture.[19]

INDUSTRY COMPETITION INTENSIFIES

This pairing of industry competitors was of some immediate concern to the Bethpage team, in part due to Lockheed's history as the prime contractor for P-3 mission system avionics upgrades via the anti–surface warfare improvement program, to their long-term relationship with the maritime patrol program office that might influence deliberations inside NAVAIR, and to General Atomics' operational success and public reputation with Predator for its lower unit flyaway cost. With the appearance of a serious competitor, however, equally rigorous competitor analysis could begin, which initially revealed two comparative weaknesses that might be exploited: Predators experienced horrific losses through operational crashes—an attrition rate that could impose a higher after-sale replacement cost to the buyer; and General Atomics needed to offer two aircraft versions—a turboprop-powered extended-range version, and a turbofan-powered extended-range version that would be able to climb beyond 50,000 feet and sustain a higher maximum speed of at least 250 knots—but at the expense of mission endurance.[20] Discriminators of speed and endurance advantage that favored the Northrop Grumman Global Hawk–based BAMS were beginning to emerge in the analysis. Going into the fall of 2003, the Navy submitted its FY2005 budget estimate to OSD for review. Funding for a BAMS engineering and manufacturing development phase was laid in, beginning with a substantial funding bump in FY2005, to achieve a BAMS initial operational capability in 2008.[21]

Northrop Grumman had good reason to pause and note with some satisfaction the achievements attained thus far. In the roughly four years since BAMS had been first brought to the Navy's attention, a new growth market had been created from whole cloth for a Northrop Grumman product line largely by Bethpage—with supporting effort by Rancho Bernardo—by dint of their trade study acumen, maritime operations analysis, systems engineering discipline, deep knowledge of Navy organization and predilections, and strategic vision and tactical execution. Despite the bureaucratic and cultural obstacles placed in front of BAMS, Northrop Grumman had attained a funded contract for two Navy Global Hawks as maritime demonstrators; had captured a funded study effort with the Australian MoD that in every sense replicated the original BAMS concept study and would shape their future buy of an AP-3C replacement manned aircraft and BAMS-like systems; had brought the Navy to the point of funding future BAMS procurement; and had created a competition for a Navy HAE UAV where none had previously existed and for which Northrop Grumman enjoyed a favorable competitive position.

Arriving at this point took no less than model leadership and committed, aggressive, and skillful execution on the part of both Northrop Grumman and certain individuals in the Navy. For the company, the arrival of David Stafford in a leadership position at Bethpage at the most opportune moment provided the right vision and encouraging mentorship that allowed key operatives—their disrupters—to formulate and put in place a winning shaping and capture campaign. Many more people played important roles at that level, but by far the principals were Frauenberger, Lapins, and Snowden. All three brought unique skill sets and experiences to bear, yet the inception of a brilliant new warfighting capability, one that was truly synergistic with the Navy's recapitalization plans, was not by itself enough to guarantee acceptance. What makes the acceptance all the more unusual is the Navy's embrace of the UAV adjunct proposition and the future years funding stream that was created to make it a reality. If, from the perspective of the business principals involved, there existed a set of marketing guidelines or qualities that by their successful application here best exemplified the gold standard for selling any military program, it could be summarized broadly as these:

1. Understand the mission: become sufficiently versed in the setting of warfighting requirements and mission-level modeling or simulation to associate a company product with qualitative improvements in Navy campaign outcomes. Using that association, set baseline expectations that reflect your company's product attributes most likely to be adopted as key performance parameters. This may be a first step to validation of a requirement to begin competing for resources but also to the shaping of competitive discriminators that will be used in later procurement competitions. For Navy requirements, it is essential to understand how they are set, which commands, laboratories, or warfare centers have a deciding input, and what the threats and technology trends are.

2. Set the objective: determine at the start what discrete end item or deliverable to the business will be achieved at the culmination of a campaign. This critical step is the determinant of success and will inform the win strategy. The objective can be winning

a competition, logging a set revenue target, or eliminating a competing program, but it must be measurable and achievable, although not necessarily viewed as "attainable" at the start of the campaign—if it was attainable, the need for a dedicated effort and resources is less compelling.

3. Know the process(es): essential to creating a Navy program out of whole cloth is an intimate familiarity with the service's program and funding priorities as expressed in its program objective memorandum. Know the procedures, rules, deliverables, specifications, and formats for how the POM will intersect with and overlap two other key DOD processes: the defense acquisition system and the joint capabilities integration and development system.[22] A comprehensive grasp of these systems, their timelines, their decision gates, and their decisionmakers and influencers will set the table for constructing a map of critical actions. Know the service chief's sentiments and priorities and, by extension, those of his or her staff deputies, in order to articulate a thematic structure or narrative that fits a POM story line and shows how your program capability aligns with department-level guidance.

4. Map the outcome: with the objective set, and with detailed knowledge of the processes, determine the critical path to achieving the end result. Getting to the objective requires navigating a set of finite, interdependent accomplishments, each occurring at some point defined by the processes referenced above. At each of those points, a mini-campaign is likely needed to organize resources and focus activity to achieve a decision by a senior official or overcome a bias by an influencer that will result in a favorable progression to a successor event. For each mini-campaign, a timeline is required to lead-turn the event with required actions.

5. Follow the money: Navy programs will always be paid for with congressional appropriations. Know the funding accounts of the money involved and how requests inside those accounts

will be scrutinized by professional staff members in Congress. Consideration of that staff scrutiny should logically begin even at the earliest inception of the requirement. The thematic narrative constructed for a POM review must hold together once the request makes its way to congressional review.

6. Engage the principals: knowing the process entails cataloguing the decision points that can make or break forward progress. At each decision point, one or only a few decisionmakers really exercise the power to advance or stop your concept from moving forward. Understand who they are and what their expectations or biases may be, and make adjustments to anticipate those. This means identifying individuals by name and engaging them directly or through surrogates to determine any predisposition on their part that can be exploited to shape a targeted message.

7. Keep your own house in the loop: back-briefing company leadership and adjacent departments and field units as needed is an absolute necessity to get buy-in and acceptance of the objective, the critical path, and the resources required. If a prototype or funded demonstration is required early on the critical path, company officials need to be anticipating that.

For the Navy, similarly decisive leadership at critical moments was provided by Adm. Tom Kilcline and especially by his BAMS requirements officer, Lt. Cdr. Mark Turner. Unquestionably, the officer who bucked conventional wisdom and defied his own community culture to seize the rare opportunity to adopt and advocate for the new technology was Capt. Alan Easterling. Certain of his vision and unbending in his persistence in the face of detractors inside the Navy, he assured the acquisition of what would become the P-8 Poseidon and the MQ-4C Triton and laid the foundation for the future of maritime patrol. By the fall of 2003, Stafford, Easterling, Turner, Kilcline, and Snowden had all moved on to different responsibilities, but each would become involved again, in different capacities, as BAMS ran into unanticipated headwinds.

CHAPTER 7
INTERREGNUM

This [GHMD] flight marks an important step for the Navy's unmanned air vehicle programs and naval aviation. For the first time, the Navy will have an unmanned system that can support the fleet from nearly anywhere it operates. The lessons to be learned from this program will benchmark future intelligence, surveillance, and reconnaissance practices for the maritime environment. Congratulations to the Navy, Air Force, and Northrop Grumman contractor team for a job well done.

—Capt. DENNIS SORENSEN, USN[1]

Legions of staff officers, civil servants, and flag officers in the Pentagon toil endlessly over spreadsheets most of the year to prepare their service budget requests for the year-after-next in a well-oiled process called planning, programming, budgeting, and execution, which superseded the program objective memorandum process. (The acronym POM, which was ingrained in the Pentagon lexicon, remained attached to the new process.) By any name, the process has form and purpose and, with periodic improvements, has proved to be an efficient means for allocating resources among competing military requirements. It reaches a crescendo after the services submit their budget estimates to the Office of the Secretary of Defense in the late summer, and a further reworking of the budget estimates is undertaken by the OSD staff to rationalize service estimates against earlier stated needs, deconflict duplicative resource requests, and force service resource requests into conformance to the secretary's strategic vision.

One mechanism by which this was accomplished for more than two decades (until 2012) was a program budget decision (PBD), an authoritative document issued under the deputy secretary's signature that adjusts program and funding profiles that are included in the president's budget submission to Congress the following February of every year. The services are accorded an opportunity to contest the PBD action with a reclama that explains and attempts to justify the request as vital to service needs. However, the PBD adjustment normally stands. In September 2003, when Pentagon analysts scrutinized a BAMS estimate from the Navy that began to steeply ramp up in FY2005 and continue well into the future years defense program—the estimated budget for five years beyond the current year—they found themselves in a conundrum. Much had occurred in the prior two years that now weighed on their consideration of the Navy BAMS estimate. Its Air Force progenitor, Global Hawk, had been rushed in its ACTD configuration into Operation Enduring Freedom in Afghanistan in 2001.

Air Force functional and resource managers were focused on wartime support, but upon return from deployment, they were met with demands for expeditious transition of Global Hawk to low-rate initial production of seven Block 10 models. Global Hawk was making this transition while concurrently still in its engineering and manufacturing development phase; it then entered into a series of spirals that required a major redesign to subsequent block models, upsizing payload and electrical power to host a multi-intelligence package, a battlefield airborne communications node package, an airborne signals intelligence payload package, and a radar technology improvement program package, each a power-draining electronic sensing black box payload. All but the radar technology improvement program were in concurrent development and integration by September 2003.[2] By then, with the added complexity and concurrency, program unit cost was racing to $50 million (from an original ACTD cost cap of $10 million) and showed no sign of abating.

A very real risk now loomed of breaching the 15 percent average unit cost growth cap set statutorily by legislation known as the Nunn-McCurdy act. If breached, the program essentially stops until the secretary of defense produces a set of certifications justifying the system's necessity. At 25 percent cost growth, the program is effectively terminated. Analysts on the secretary's Pentagon staff were mindful that the Air Force's Global Hawk estimates were most probably understated and would result in further difficulties. At the same moment, the Navy was seeking a seven-fold increase in research and development funds going from FY2004 to FY2005, followed by a doubling of its FY2005 request in FY2006, required to provide for a transition of MMA to a system development and demonstration phase.[3] From the perspective of the Pentagon analysts attempting to make ends meet across a department-wide budget, Air Force Global Hawk needed a boost, and the Navy appeared more than willing to sacrifice BAMS to ensure that the MMA estimate was preserved.

In a single stroke, the Pentagon staff sent BAMS into a tailspin: draft PBD number 707 and an accompanying budget change proposal removed $185 million from the Navy funding line for BAMS in FY2005 and transferred most of that to Air Force Global Hawk in early November 2003.[4] The cut to BAMS funds was sufficiently sweeping to eat into funds planned for GHMD execution as well. Wrangling between the Navy and Pentagon civilians continued into December, when the Navy accepted the inevitable. During those fall months, Northrop Grumman's BAMS team in Bethpage, aware that BAMS funding was in extremis at the senior levels of the Pentagon, went into crisis mode.

Under Lapins's direction, with guidance from Teel, the team prepared an unsolicited offering that they began calling a rapid response proposal. The proposal was intended to preserve what were believed to be Navy priorities (as best understood at the time without access to the classified operational requirements document annex) by redistributing development and production efforts to fit within what now appeared to be available funding. As a highest priority, the rapid response proposal advocated preserving the GHMD aircraft and associated systems to enable an early operational capability in FY2006. An earlier declared operational capability would have placed the system in the fleet user's hands to wring it out and confirm its utility as the program struggled to regain its footing—hopefully producing useful testimonials for later advocacy briefs. The rapid response proposal further proposed delivery of ten production aircraft over the next three years; intended to regain some stability and achieve the formal initial operational capability in FY2008; proposed fielding a mission control segment for near-real-time tactical exploitation, enabling aircraft carriers at sea to receive downloaded imagery and data by FY2008, with remote control of BAMS sensors and flight path from the carrier by FY2010; and, finally, conducting full and open sensor and subsystem competition immediately upon contract award to Northrop Grumman.

The Bethpage BAMS team position, which was the preferred approach to recommend to the Navy in view of the funding pressure, would rest on bifurcating the system development and demonstration phase into two spiral developments—the first producing a minimum system capability that was partially compliant for initial operational capability in FY2008 (initial operational capability defined as early capability plus one forward-based unit to support one persistent ISR orbit), and a second spiral to be fully threshold compliant with the operational requirements document by FY2010, with retrofit of previously delivered units to the full threshold capability. Both spirals would be governed by cost-as-independent-variable strictures and, when imposed with rigor, portended promise of holding the line on cost stability and predictability.[5]

The Navy resubmitted its estimate that acknowledged and accounted for the reduction in the BAMS funding line to the Pentagon civilians on December 22. By January 13, 2004, the other stakeholder in the process, the Air Force, acted in response to the unfolding budget drama by issuing a stop work order on the GHMD effort. In so far as the Air Force knew officially, the Navy would no longer be a paying partner in the Global Hawk Block 10 production, and their action to cease any activity on the two Navy aircraft seemed prudent. This deepening quandary that now put GHMD potentially at risk impelled Dr. Ronald Sugar, Northrop Grumman chief executive officer, to weigh in. After visits with the deputy assistant secretary of the Navy for air programs, the Secretary of the Navy, and the under secretary of defense for acquisition technology and logistics in the span of two days in late January, things appeared to recover somewhat for GHMD, with BAMS still headed for open competition.

Dr. Sugar's representation on behalf of the BAMS team was helpful, if only in making a strong showing for the company's commitment and preferences, but more so in adding to a growing sentiment among Navy principals that moving forward on GHMD was important to informing a BAMS future and to augmenting MMA. After meeting with the Navy

In early January, the US Navy announced that it would remove as much as $500 million from the $2 billion Broad Area Maritime Surveillance (BAMS) program in its 2005-2009 spending plans. At press time, the BAMS program office was still planning to release a draft request for proposals in late January.

—"U.S. Navy Briefs," *Journal of Electronic Defense*, February 1, 2004

deputy assistant secretary and the CNO's staff, the Navy program manager then acted on his direction to issue a notional plan for competition that included a crash schedule to make a system development and demonstration award inside the coming twelve months. It would take much longer to regain the initial BAMS momentum established in 2001–2.

In an ironic reverse, just as BAMS was slipping in the Navy, Global Hawk's reputation and progress were solidifying in the Air Force. Coming off Operation Iraqi Freedom, when U.S. forces were trudging toward Baghdad, an early ACTD Global Hawk aircraft on deployment by the Air Force to the Middle East lit up "thirteen surface-to-air missile batteries, fifty SAM launchers, seventy missile transporters, and had imaged 300 Iraqi tanks" that would otherwise have been obscured by a persistent sand storm enveloping U.S. troops; "it was at that time, early 2004, that the Global Hawk program became a genuine part of the Air Force plan."[6] Now, Air Force leaders were looking at the Navy's plan to compete BAMS with some chagrin. The addition of the Navy to the Global Hawk program, despite having a different basing scheme, operating rhythm, and emerging internal configuration differences, still offered a bright opportunity to double the production run and gain significant economies in both manufacturing and post-delivery support. That could be lost if a Navy competition delayed Navy selection of Global Hawk past a point of synchronizing with Air Force production or went to a bidder other than Northrop Grumman that would introduce a system into the Navy inventory that was not compatible with the Air Force Global Hawk. Air Force leadership felt no reluctance about engaging directly with the Navy to make the case that competition was not in either service's best interest in the longer term. Air Force secretary James Roche and chief of staff Gen. James Jumper directly lobbied their Navy counterparts—all to no avail: "In twin letters to Navy Secretary Gordon England and Chief of Naval

Operations Admiral Vern Clark, the Air Force leaders said a joint Global Hawk fleet would produce savings."[7]

The Navy's course was set, and the leadership would not be dissuaded. One can postulate that senior Navy leaders believed competition forced Northrop Grumman to become more cost-competitive on the one hand and averted the disruption of industry protests from potential competitors on the other. Or they may have entertained the notion that a competitive challenger to Global Hawk might emerge that could come close to satisfying the conclusions of the BAMS concept. Then, too, competition allowed the BAMS schedule to stretch out considerably, giving MMA breathing space to establish a solid footing.

For the Air Force, this entreaty to Navy leadership to join the Global Hawk program might be seen in hindsight as a last-ditch effort to avert a looming programmatic crisis or at least mitigate the untoward effects. Following years of underfunding, tepid support from Air Force leadership, excessive concurrency while developing a fundamentally new B model, and over-stretched deployment into combat theaters, Global Hawk was headed at breakneck speed into the snare of a Nunn-McCurdy breach.[8] In mid-April, the Air Force notified Congress that the procurement unit cost of Global Hawk had increased 18 percent over the baseline estimate. The Nunn-McCurdy act established limits on cost growth that, if exceeded, require congressional notification. Two measures are pertinent: program acquisition unit cost, or the per-unit produced cost of developing and procuring the program, and the average procurement unit cost, or the program's procurement cost per unit produced. These measures are gauged in relation to a program's originally approved milestone B factors and current baseline if the program has been rebaselined. In this instance, the threshold that had been exceeded was 15 percent, termed a significant breach. A critical breach—where cost growth exceeds 25 percent and requires recertification of the program by the secretary of defense—would be triggered by the end of the year. But now, in mid-spring, there was enough uncertainty surrounding the Global Hawk program that the Navy, involved in a minor way with GHMD, was satisfied that opening BAMS to industry competition was its best course. The delay of any BAMS procurement decision now had the added justification of allowing the Air Force time to sort out its difficulties and put in place recovery steps before a BAMS competition led the Navy back to Global Hawk.

Whatever justification prevailed for delaying BAMS and moving forward with the manned aircraft, MMA source selection was decided and announced by June 15, 2004. Boeing was presented a cost-plus-award-fee contract for $3.8 billion to take its militarized 737 large passenger jet derivative into a system development and demonstration phase. The Navy assessed that for the development phase, the Boeing proposal was slightly less expensive than the Lockheed Martin proposal. Because this evaluation was a best value determination, other factors were accorded some weight; specifically, the Navy assessed that, owing to 737 airframe experience, the Boeing offering held the potential to deliver as much as a year sooner. If realized, that potential could mean cost avoidance on fleet modernization of existing P-3s.[9] The selection of the Boeing commercial jet may have been a foregone outcome, given the Secretary of the Navy's pronouncement three years earlier when approving the program element stand-up for MMA: "MMA would be a commercial airliner!"[10] There were pockets of skepticism in the maritime patrol community that doubted the ability of the twin-engine jet to perform long missions low over the water.

The Navy Secretary's prejudicial declaration notwithstanding, Boeing showed greater focus, intensity, and commitment to winning MMA over the next three years than did its competitors by addressing the doubts directly. Boeing's marketing team launched its own socialization campaign in November 2003, taking a 737 on a whistle-stop tour of maritime patrol bases for familiarization (and reassurance) flights by resident naval aviators.[11] Boeing outfitted a rolling tractor-trailer rig as the proposed interior of the 737 MMA to demonstrate operator-machine interface and functionality considerations to naval flight officer mission commanders at multiple venues. Lockheed Martin, perhaps resigned by that point to a second-place finish, pointed to its former P-3 production program days and ultimately submitted a lackluster (compared to Boeing's) proposal.[12]

Once again, Lockheed Martin surrendered a franchise incumbency, perhaps its last substantial position in the manned maritime patrol aircraft market. But the Navy's announced intent three months before the

MMA decision to open BAMS to competition gave Lockheed Martin new entrée into the maritime patrol arena. Lockheed Martin and Predator UAV partner General Atomics embarked on a series of demonstrations to underscore its UAV's utility in maritime surveillance. In quick succession that summer, the Lockheed Martin/General Atomics team pushed their Predator derivative, Mariner, through its paces in a data collection trial off San Diego and in similar tests with the Coast Guard off Alaska and the Canadian government off the Atlantic coast. Not demonstrated that summer was anything approaching 360-degree ocean coverage with either of their advertised candidate radar systems, the Telephonics radar or Raytheon SeaVue radar, nor the range and altitude commensurate with early BAMS concept exploration study recommendations.

Still mindful of Philip Teel's exhortations to exploit the solid BAMS position to find a way back into the MMA program, Snowden assumed the lead by seeking out for Lapins and himself an appointment with the winning Boeing team for initial conversations about teaming possibilities. They constructed a short list of objectives and drafted talking points to guide the interaction with Tim Norgart and possibly others from Boeing's MMA capture team in the hope of directing the conversation toward some collaborative endeavor, probably a joint marketing strategy that further developed the notion of operational connectivity between MMA and BAMS and the increased mission utility thereby obtained. When they met, discussions were cordial, but the results were indefinite. It was apparent that the Boeing team was holding their cards close to the vest, most likely anticipating that they needed to pause and savor the implications of their MMA win; that Northrop Grumman, in their view, might not ultimately win BAMS in an open competition—much depended on how the competition was run and who competed; that Boeing might have a play in the BAMS competition that would need to be further researched; and that BAMS was a distraction, a possible draw on Navy budget obligation authority that could undermine MMA, and Boeing's best strategy might be to delay and ultimately kill BAMS to preserve higher numbers of MMA production and inventory.

Given the outcome of those discussions, Northrop Grumman's BAMS team turned its efforts wholly to preparing for a BAMS competition, with a solicitation to industry for proposals expected within weeks if not a few months. Teel resolutely believed that BAMS was the key to reentry into the MMA competitive field, despite a prevailing industry view that BAMS was inimical to stabilizing the MMA program. In progress reviews held in the Bethpage conference room, his comments suggested to those present that MMA was of at least equal importance to BAMS and, accordingly, he assigned Lapins the additional goal of shaping a strategy to make that possible.

The outline of the MMA plan—which still included a possible partnership with the BAE Corporation on a resurrected British Nimrod patrol plane—was taken to a sector-level staff review at the headquarters in Las Colinas, Texas, with sector head Ralph Crosby. Ralph considered and then rejected any MMA distraction, encouraging Teel to "forget MMA and focus on BAMS."[13] He went on to acknowledge the adversarial atmosphere between Bethpage and the UMS division at Rancho Bernardo over BAMS and to express his frustration with the inability of his two division leaders to arrive at some accommodation. Exasperated, he pulled out a blank legal pad and drew a Venn diagram to reinforce to Teel and his West Coast counterpart the area of common ground; both were admonished to negotiate a resolution for more harmonious operations for BAMS.

Coming out of that sidebar discussion, Teel assigned Lapins the task of drafting a team charter to resolve major differences in the most equitable fashion possible. Perhaps indicative of the hardened positions already staked out by both divisions, neither side took up the draft charter for execution for at least another year. Conditions worsened as the two divisions sparred. Crosby might have missed an opportunity at this point for leadership oversight by not demanding closure from his two division heads. At UMS, indignation at taking a subordinate role to the Bethpage group for a product in its portfolio, Global Hawk, had generated no small amount of acrimony over the course of four years.

The Bethpage group met that sentiment with a healthy disdain for UMS's alleged Air Force–centric bias that would undo the program if allowed to prevail. That bias was manifested in UMS's singular pursuit of total design commonality of the Navy BAMS with the in-production

Air Force configuration, essentially dropping 360-degree radar coverage, MMA connectivity, and other distinctly maritime attributes from the original Bethpage-drawn BAMS concept study recommendations—attributes strongly desired by the Navy. Through the summer, Lapins, Frauenberger, Snowden, and others who had been associated with BAMS in Bethpage were detailed to support the proposal operation in Rancho Bernardo, under the direction of a proposal manager and volume leaders appointed from UMS and the sector staff. It seemed to Snowden and Lapins that while Teel was overly focused on MMA workshare, UMS in Rancho Bernardo had successfully lobbied the senior sector staff and reclaimed its role in leading the proposal effort. Lapins was assigned as deputy proposal lead, subordinate to a UMS appointee; Snowden was assigned as the business development lead for the proposal and associated capture campaign, subordinate to the vice president acting as dual-hatted proposal and capture lead; and Frauenberger was assigned to support the mission analysis and configuration work, subordinated to another supervisor at UMS.

This arrangement was announced to be in effect for the proposal only but was a foretelling of decisions that might have been in the works. The tone for the Bethpage-Rancho commingled proposal team was set by the introduction of the senior red team reviewers. Heading that team was Ollie Boileau, brought out of retirement to impose summary discipline on the review process. This was the same Northrop officer who had wreaked resentment and ill will at Bethpage ten years before in bringing the former Grumman work force to heel in the wake of the acquisition. For Bethpage participants with long memories, this was an especially inauspicious beginning. Most had observed on at least one occasion Lapins and a UMS counterpart squaring off on opposite sides of a large conference table, unable to reach a consensus over details of proposal volume construction. The enmity was in plain sight.

The dual-hatted proposal and capture manager, a vice president detailed from his advanced programs responsibility on the sector staff in El Segundo, California, understood he had a problem and that it had everything to do with East Coast and West Coast divisions, with each asserting its right to manage not just the proposal but more importantly the Navy

BAMS program that would hopefully follow a winning submission. This reemergence of deep-seated cultural antagonisms that had smoldered for many years likely was provoked by the appearance of Boileau. From an East Coast perspective, it was the contempt for the Northrop (and now Ryan) upstarts and their absorption with the elegance of their technological solution, supposedly affirmed by their Air Force customer. From the West Coast view, it was disdain for the singular focus of the East Coast division on catering to the Navy customer and the supposed long family "in-breeding" that stunted technological vision in Bethpage.

The proposal and capture vice president spent the better part of the summer sampling opinions from company officers inside and outside those divisions and outside the sector to cull out those practices and individuals contributing most to the dysfunction. He constructed and administered a survey to those division and sector officers that rated the two divisions on relative strength in systems engineering skill base, avionics and airframe design, operations analysis and simulation, experience with the Navy customer, past performance, production base, and other areas deemed vital to program execution. The survey tipped the balance to Bethpage, but he chose to keep the results embargoed. If there was an error in judgment, a deficit in leadership, it emerged again here with the vice president's failure to act decisively and to communicate to all stakeholders the action taken and the motivating rationale. Corporate leadership theorists point out that "the first step in changing culture is communication and information sharing. . . . The reasons and logic underlying the need for change must be complete, unambiguous, and compelling. . . . The data supporting and justifying change must reach the right people."[14]

For this sector vice president, several options may have been within reach for modifying behavior on the proposal team rather than living with irreconcilable cultures and hoping for the best outcome. In his work on strategy and culture, Lawrence Hrebiniak enjoins managers to resist attempts to change a culture's norms or credos and instead target incentives and organizational structure. On this proposal team, incentives were implied but not tangible: a winning proposal would increase the prospects for more permanent work on the program, nothing more.

Bethpage representation on the proposal team was barely visible. Those detailed to Rancho Bernardo were given roles subordinate to UMS proposal book and volume managers. That any Bethpage people were at all evident seems to have been by design to ameliorate Navy concern about losing trusted counterparts.

A better approach—in hindsight—may have been to anchor the proposal effort in Bethpage and detail a few key proposal writers and technical experts from Rancho Bernardo in subordinate roles to Bethpage volume leaders. This action could have had the more desired effect of acculturating the UMS people in the inclinations and idiosyncrasies of a Navy customer that Bethpage had clearly won over as well as reassuring Navy counterparts that the proposal would be prepared by people with whom they had worked intimately for more than four years. It would also be more in conformance with the survey taken by the detailed vice president and capture manager. Nonetheless, the proposal activity that summer of 2004—despite the nearly debilitating internal Northrop Grumman cultural maladjustment—would be rendered moot before a solicitation to industry would ever be issued, stopped dead in the water by Navy budget action that essentially gutted BAMS for the foreseeable future.

In an ever-constricting budget environment for naval aviation, the Navy's abrupt excising of BAMS funding solved several immediate problems. In the first instance, the first of two GHMD aircraft would soon make its appearance, and there was rising concern about the concurrency of a GHMD demonstration program and the BAMS program that it was intended to inform. Indeed, the National Academy of Sciences, under a Navy study contract to assess the state of UAV adoption, reported that "the road ahead seems unclear for the long-dwell, standoff Intelligence,

The Navy's FY06 to FY11 budget plan proposes a three-year delay to the Broad Area Maritime Surveillance (BAMS) unmanned aerial vehicle (UAV) program—a move that critics say indicates a lack of commitment to UAV development from the service. A three-year delay could shift initial operational capability as far out as FY13; it is now set for three years earlier. Before the delay, the Navy had planned to field an early operational capability for BAMS in FY08.

—AMY BUTLER, "Navy Proposes Delay to BAMS UAV Effort in FY06 Budget," *Defense Daily*, September 17, 2004

Surveillance, and Reconnaissance, or ISR system. . . . The committee noted the near-concurrency of the GHMD and contract award for the BAMS UAV, and thus it remains concerned that lessons from the Global Hawk demonstration might not be reflected in the BAMS program."[15]

Second, naval aviation's first priority for maritime patrol was recapitalization of the manned P-3 aircraft. Now that it was under way, the maritime patrol community and its representatives on the CNO's staff and in NAVAIR were concerned about the near-concurrency of the manned and the unmanned aircraft development and their follow-on production. In its unitary logic, the Navy bureaucracy could not conceptually deal with more than a single new-start aircraft development and procurement within a single mission area that rested somewhere in the hierarchy of needs below full-rate production of the F/A-18E/F, transition from development to production for the V-22, and joint strike fighter development that appeared increasingly problematic. For the Air Force, the Navy's action signaled a lost opportunity to gain economies of scale that would have been achieved with the timely addition of Navy resources. And Navy participation could promote the allure of "jointness" added to the stability of the Air Force Global Hawk program.

The outcome for Northrop Grumman was a similar loss of production efficiencies that might have been obtained if production of Global Hawks for Navy maritime ISR was synchronized with Air Force Global Hawk production. The proposal team that had been established in Rancho Bernardo from members from the two sector divisions and that struggled through the summer of 2004 with internecine rivalry was disbanded. All focus on the Navy Global Hawk side of the ledger now focused on GHMD and ensuring that it met Navy expectations. Navy satisfaction with GHMD and Northrop Grumman's support of the aircraft system and its planned experimentation would be key to developing a compelling win theme when and if the BAMS competition restarted. For the moment, the future of BAMS depended on the future of GHMD—and the UMS division in Rancho Bernardo held that contract.

Captain Easterling left the maritime patrol program office shortly after his groundbreaking work resulted in the award of the BAMS concept study contract. His last active-duty assignment was in a newly created office in the engineering directorate of NAVAIR, where he developed an underlying architecture for a concept that had come to dominate PowerPoint presentations both on the CNO's staff and in NAVAIR. The concept, FORCEnet, had been given its earliest practical description in the work of the CNO's Strategic Studies Group XX in 2001 and was refined the following year. In those reports, FORCEnet was defined as "the operational construct and architectural framework for Naval Warfare in the Information Age which integrates warriors, sensors, networks, command and control, platforms and weapons into a networked, distributed combat force, scalable across the spectrum of conflict from seabed to space and from sea to land . . . relying on near-instantaneous collection, analysis, and dissemination of information over seamless communication paths coupled with computer-driven decision aids to unify the perception of the battle space in the future."[16]

As laborious as that definition may sound to the uninitiated, it was just as difficult for the experts to reach any consensus on equipment, linkages, protocols, resources, even laboratory-bound breadboards or brassboards that made the concept seem real and effective. This issue had been incubating for several years since the strategic study group reports when Easterling arrived at his last assignment. Here he was to try and put form and substance to the cartoon lightning bolts symbolizing FORCEnet in so many briefings. His progress in that quest was measured, but it equipped him with the intellectual insights that would be used to reinvigorate BAMS in his next assignment after retiring from active duty.

Easterling entered the Applied Physics Lab (APL) at the Johns Hopkins University as a systems architect. At that same time, a new methodology was gaining currency for preparing the Navy POM. More emphasis was now being placed on the intersection of warfighting capabilities, where integration into mission architectures allowed redundancies to be revealed and resourcing priorities to be more clearly drawn. To aid in this process, Pentagon leaders debuted large-scale scenarios called major contingency operations as the context for developing operational and tactical summaries for simulations. These simulation results, in turn, provided the basis for mission effectiveness and justification for programming resources in the POM request to OSD. Partly in response to this new methodology, APL assigned Easterling to assist the Navy UAV program office in making order out of a dismal situation and to adopt a more analytical approach to gauging BAMS effectiveness. The start-stop nature of program activity, the depth of the competitive field, and the lack of an analytical means to level the field and derive solid measures of utility in maritime ISR were gaining only abysmal reviews in senior secretary-level briefings. From his APL vantage point, and with his program office relationships still fresh, Easterling urged the UAV program office to consider replicating what had been done in the early stages of MMA by getting industry involved in doing the trade studies against a set of mandated operational situations and tactical situations. This evolved into a more formal structure for industry involvement that bore the title persistent unmanned maritime airborne surveillance (PUMAS). That cadenced acronym was the creation of Paul Achille, then civilian deputy in the program office.

The PUMAS solicitation was to have the aerospace industry provide outside independent assessments of the utility of manned and unmanned maritime ISR systems, circa 2013, by developing recommendations for an affordable family of interconnected systems. Further, it was expected that the PUMAS studies would identify the key performance parameters for the BAMS system specifically. Easterling would oversee the day-to-day conduct of the studies and provide course corrections as needed. Capt. Bob Dishman came aboard as the program office deputy in charge of BAMS, with the concurrent arrival of Tom Garrett from the NAVAIR program management competency as lead engineer.[17]

PUMAS would draw intense interest from the BAMS competitive field, such as it existed in the buildup to the aborted 2004 BAMS request for proposal (RFP). The principal competitors had not varied since the BAMS industry day in December 2002; they were still Northrop Grumman with its Global Hawk derivative, and Lockheed Martin teamed with General Atomics for the Predator derivative, now designated Mariner. Boeing's "Phantom Works" was encouraged at this point to get in the mix with a notional and somewhat radical design they had been pursuing called Sensorcraft—a kind of blended-wing concept. Sensorcraft, though innovative, would not gain traction—probably due to being too unconventional. Boeing ultimately dropped it in favor of a more traditional concept and industry teaming, choosing to align with Gulfstream to offer an unmanned G-550 business jet.[18]

The central theme of the PUMAS study would be to identify a family of systems that would be the most capable and cost-effective in conducting the Navy's persistent ISR mission by 2013. Reduced to its essential purpose, PUMAS was to rationalize the growing genus of like competitors and extract only those attributes that favorably complemented or augmented a 737 passenger aircraft derivative in the manned aircraft role. The Navy set certain conditions: the family was identified as MMA (still a notional system, as a winner had not been yet announced), the aerial common sensor (the as-yet-undefined system to replace the EP-3), and unmanned systems that included a notional BAMS and surface ship–based UAV. A great deal of emphasis was placed on a basic BAMS goal of twenty-four-hour time on station at two thousand nautical miles, but even that could be traded off to hold costs down. As the PUMAS RFP drew near, a critical decision loomed for Northrop Grumman. The experience of the previous summer of trying to blend a proposal team of Rancho Bernardo and Bethpage people on the West Coast under UMS leadership had proved problematic at best. That incarnation lacked cohesion internally and failed to garner Navy trust and enthusiasm. The leadership failure of the previous summer—failure to announce a decision during proposal effort of the previous summer or even reveal the existence of the analysis to the team—was rectified this time, mainly by external forces.

Before an RFP was released, Easterling contacted Frauenberger to learn if Northrop Grumman intended to respond, taking care to emphasize that if so, the Applied Physics Lab—and by implication, the Navy— would view with favor a Bethpage-led effort that involved Frauenberger. That input, although received second-hand from the Navy customer, was enough to sway a more forceful decision this time. PUMAS was deemed a strategic must-win sector-level effort, to be overseen by the same advanced programs vice president. However, this time, the Bethpage division would lead from Bethpage. Frauenberger was designated the study manager, much to the satisfaction of Easterling and his Navy sponsors but to the consternation of the UMS people on the West Coast.

Naval Air Systems Command has issued contracts to four defense contractors to study Persistent Unmanned Maritime Airborne Surveillance (PUMAS). The Navy commissioned the studies to "aid development of a system-level alternative that can provide optimal unmanned persistent maritime ISR. . . . Boeing Integrated Defense Systems, General Dynamics Advanced Information Systems, Lockheed Martin Maritime Systems & Sensors Tactical Systems, and Northrop Grumman Integrated Systems were awarded contracts of approximately $1 million for initial five-month studies. After review, the Navy will down-select contractors for an additional seven months of study under contracts each worth approximately $3 million."

—RICHARD BURGESS, "Companies Studying Persistent Unmanned Maritime ISR for Navy," *Sea Power*, October 1, 2005

To galvanize the assembled PUMAS proposal team, the vice president in his kick-off address noted that for this effort, the "corporate implications were very big. . . . failure is not an option!"[19] The core Bethpage team would be augmented with representation from UMS at Rancho Bernardo (key technical experts on the aircraft and ground control segment), from the former PRB subsidiary at Patuxent River (prime contractor to the Navy for tactical support centers), from Northrop Grumman electronic systems sector (for radar), from Northrop Grumman mission systems sector (for command and control), and from L3 Com (for key sensor input).

In order to promote frequent dialogue with the industry teams, Easterling laid in three technical coordination meetings for the first phase of PUMAS, spaced throughout the period of performance and hosted at the contractors' facilities to give them the widest latitude in what they presented and could display in their labs and simulation control rooms that was directly relevant to their work on PUMAS. By the second technical coordination meeting between the Navy and Northrop Grumman, back-channel feedback from the Navy revealed growing dissatisfaction with some elements of the Northrop Grumman team, especially UMS positions being briefed in the meetings in which it was apparent that they were trying to "force-fit the Air Force Global Hawk vehicle and systems on a maritime mission"[20] rather than keeping the trade space open for maritime operational and tactical situations.

The Navy's discomfort was expressed openly and directly when a UMS quote was read to the Northrop Grumman PUMAS team leadership at the meeting outbrief. "It's good enough for the Air Force, so why isn't it good enough for the Navy?"[21] was the expression heard most often and quoted back to the assembled team. Much of the Navy's heartburn with this intrateam lobbying centered on radar configuration. With the Northrop Grumman electronic systems sector on board as part of its PUMAS study team, an early advantage in maritime surveillance was indicated in the analysis for a derivative of their APY-6 radar with 360-degree electronic scan. This comported well with Bethpage's original recommendation in the BAMS concept study from five years earlier, as well as with the Australian maritime airborne study that followed in 2003. The essential benefits of a 360-degree scan versus a side-looking scan had been clearly borne out in the Australian study, and it was taken as an article of faith by the Bethpage group that the matter was settled. The UMS Rancho Bernardo representatives on the team were vocal advocates for the Raytheon side-scanning system that was part of the Air Force Global Hawk vehicle. The UMS position was understandable; a common sensor configuration with the Air Force would minimize schedule and cost risk in development and promote economies in production. Their insensitivity to the Navy's perspective and preferences only served to foment worsening tension between Northrop Grumman's East Coast and West Coast study team members.

As the trade studies progressed, trends could be discerned in each contractor's briefings. Boeing seemed to be pointing to speed and time to station as a greater measure of effectiveness over persistence. They seemed not to understand that speed contributed to reconnaissance efficiency. But in the end, the most fundamental measure of effectiveness for BAMS was dwell. Likewise, the Lockheed Martin/General Atomics team seemed to be skewing their outcome to low-speed advantage, searching for a metric that would promote Predator vehicle advantage. Northrop Grumman team members were incredulous when hints filtered out about the competitors' heavy thumb on the scale, suspecting different teams were working to different guidance. When Northrop Grumman members made that observation and inquired about their rationale, APL and Navy evaluators provided an imprecise answer that hinted at management direction in those companies driving their results that way in an apparent attempt to bias the government evaluators. Of the three contractor teams engaged in PUMAS studies, APL and Navy evaluators gave the Northrop Grumman team strongest marks for holding to as agnostic a view as possible, treating their notional candidate systems as equally as possible throughout their trades. This made them stand out as the contractor teams drew closer to completion of the performance period.[22]

A very critical part of the Northrop Grumman study, and one not called out specifically in the contract statement of work, was a detailed "ghosting" of the General Atomics Predator B–based Mariner. Ghosting involved identifying the competitor's shortcomings and highlighting them indirectly—without naming the competitor—by drawing distinctions between those shortcomings and advantages obtained with the preferred product relative to the desired attributes. Making these assertions credible required a detailed analysis and understanding of the competitors' strengths and shortcomings. The Northrop Grumman team assembled a focus group to do a thorough red team review of the competitors as the starting point for the ghosting. The team's charter was to assume the role of the competitor, conducting parallel analysis of the competitor system to tease out performance deficiencies. In their PUMAS study report, a graphic representation of analysis results was developed that measured multiple attributes on vectors emanating from a center, effectively producing a carpet plot of "goodness" for key maritime airborne UAV ISR elements. This spider chart—so named because it resembled a spider's web to the untrained eye—revealed that longer, faster, and higher, as measured by aircraft speed, surveillance persistence, operating altitude, and growth potential, were hallmarks of effective UAV ISR.

By implication, then, the competitors scored lower (i.e., less capable) in those areas along those vectors. Of particular importance to the Applied Physics Lab and Navy acceptance of these assertions was the support provided by Frauenberger and his team to APL and the Navy in developing a template and supporting model for effective time on station (ETOS) that became a primary and standardized comparison matrix between contractor teams.[23] In February 2006, with the first phase of PUMAS complete, the Navy took only two contractor teams into a phase two, which would last another seven months. Northrop Grumman and the team of Lockheed Martin and General Atomics would refine system concepts, sharpen cost estimates, and identify risks in phase two, gathering with the evaluators for another four technical coordination meetings before the study was complete.

By the end of phase two, Northrop Grumman's PUMAS results, largely accepted by the Navy as the most straightforward and unbiased of the industry reports, documented a solidly compelling case for a modified Global Hawk in the role of BAMS. The foresight and care taken by the Northrop Grumman PUMAS team to analyze and ghost the competition generated undeniable discriminators that could be reintroduced when and if the formal competition for BAMS ever regained its footing. The final report revealed that surveillance persistence was the key measure of merit for airborne maritime ISR within the conditions set by the Applied Physics Lab and the Navy for the PUMAS analysis. This was an undeniable affirmation of Northrop Grumman's assertion in their original concept study that a high-altitude, long-endurance vehicle provided the optimum solution. The team determined that surveillance coverage from on-station orbits, attainable only in a Global Hawk class of vehicle, could reach an ETOS of 85 percent. That measure would later become a threshold on which Northrop Grumman competitors would be scored.

Behind surveillance persistence, in descending order of importance, the team cited a set of mutually supporting factors: higher speed was the key to lower transit time to get to station; higher operating altitude provided the widest possible field of regard for the sensor apertures, the ability to fly above strong jet stream winds, which could worsen a vehicle's time and fuel available to attain on-station position, would offer greater sanctuary from surface-to-air threats and provide more efficient turbine engine operation for lowered fuel costs; ample engine and electrical power afforded growth potential to host a varied set of sensors that addressed ISR gaps revealed in the operational and tactical situations; and the aforementioned persistence, speed, and altitude advantages combined to yield greater flexibility in basing the air vehicle at any number of maritime patrol sites or forward operating bases.

It is worth noting that from the inception of BAMS in 1999 through the original BAMS concept study completed at the close of 2000, very little changed in system design, ISR performance, and notional basing laydown. Certainly, this was due in part to the maturity of the Air Force Global Hawk from which the BAMS gestated. In greater part, it was because of the innovation, competence, and discipline of the Bethpage BAMS team. The one constant through that evolution was the presence of Frauenberger as not just the study manager but also the disrupter. He was unwavering in carrying forward his radically new vision and imposing analytical rigor on his team, which then mentored his Navy and APL counterparts, nurtured those relationships to an exceedingly high level of trust, and remained true to the original vision of transfiguring an Air Force Global Hawk to a Navy mission with Navy sensor payload. To be sure, Frauenberger operated in league with other disrupters—Lapins, Easterling, and Snowden—and drew on their particular expertise at appropriate moments to maintain headway. But it was largely his efforts and the milieu in which he worked— the long-established pattern of new design concept iteration bounded by a deeply rooted naval aviation brand identity.

In March 2006 at the mid-point of the two PUMAS study phases, the first Global Hawk Maritime Demonstrator aircraft arrived at Naval Air Station Patuxent River to begin a demanding schedule of familiarization and experimentation. Piggybacked on the existing Air Force Global Hawk contract, GHMD brought with it a management structure dominated by the UMS division from Rancho Bernardo and headed by Carl Johnson, bringing his prior experience as vice president of the Air Force Global Hawk program. Johnson's earlier tenure on the Northrop B-2 program gave him the necessary stature and standing with the sector leadership to ensure his internal funding and personnel assignment needs were given full accommodation to make GHMD a constructive influence on a future BAMS competition. In a move for greater administrative and resourcing efficiency, the sector now collapsed its divisional structure from two West Coast divisions (unmanned systems in Rancho Bernardo and air combat systems in El Segundo) and two East Coast divisions (airborne early warning and electronic warfare in Bethpage), and the battle management systems in Melbourne, Florida, to a single Western region and Eastern region.

With Philip Teel's departure from Bethpage, the sector leadership in El Segundo figured to restore hegemony over the sometimes recalcitrant

Bethpage and Melbourne operations by importing another West Coast B-2 program alumnus, Duke Dufresne, to oversee the consolidation of the former Grumman units. Dufresne arrived without much introduction, and his first offsite to gather and inspire the Bethpage and Melbourne senior leadership failed to persuade many that he brought a personal enthusiasm for taking the eastern region in a bold direction or even for living on the East Coast. His announced vision for investing in National Aeronautics and Space Administration programs was not received warmly by a leadership team that had inherited the mantle of their legacy Navy prime forebears, despite Grumman's much earlier role in the Apollo program. Suspicions (later confirmed) were voiced privately that he had been parked in Bethpage pending later reassignment to another West Coast position. But for the moment, with the completion of the PUMAS study for the Navy in Bethpage in August, the sector focus reverted to GHMD. With GHMD contracting placed under the Air Force Global Hawk program and Carl Johnson anointed as GHMD manager, the leadership of BAMS for the Navy also defaulted to Johnson, now reporting to Dufresne in that role. Any concerns from the West Coast sector or division headquarters about Bethpage reasserting its individualism were allayed.

At this moment, however, Dufresne received a telling personal communication from the Navy that was cause for reappraising his view of Bethpage's competence on the eve of his return to the sector staff on the West Coast. In a letter from UAV program manager Capt. Paul S. Morgan, USN, commending the Bethpage PUMAS team for its study output, Dufresne was told:

> Under the leadership of Joe Garone [who relieved Lapins as head of the Bethpage ATDC] and Howard Frauenberger, the Advanced Capabilities Development organization assembled a crack, cross-disciplinary team. . . . From the first Technical Coordination Meeting, the Northrop Grumman comprehension of task was as evident as the commitment to task. . . . The Northrop Grumman PUMAS Team was thoroughly responsive, professional and at every juncture willing to take the extra step to ensure the Navy was receiving timely and useful product. . . . Interaction between the Navy and

the Northrop Grumman PUMAS Team was characterized by candor, mutual learning, and constructive critique. . . . The Northrop Grumman assessment was thorough, objective, and grounded in physics and engineering based analysis. . . . The Northrop Grumman PUMAS Team exceeded Navy expectations![24]

A hand-penned personal salutation from Captain Morgan on the official letterhead noted, "Duke: Your folks have been outstanding!"

The Navy's unmanned aircraft programs had now matured to the point that a reorganization on the Navy side was necessary. Whereas most of the UAVs of all sizes and intended missions had resided for several years in a single program office, programs were now parsed among a new set of program management offices aligned to capabilities. BAMS and GHMD were now hosted in a separate program office, which was given the title persistent maritime unmanned aircraft systems. This new program office retained program oversight, but operational employment of GHMD in exercises fell to Air Test and Evaluation Squadron Two Zero (VX-20) residing at Patuxent River. This had the benefit of collocating GHMD airborne evaluation in the same test squadron that inherited the manned maritime patrol aircraft, MMA. For exercise and experimentation play, the Navy Warfare Development Command assumed the lead role for planning.

GHMD aircraft were drawn from the initial post-ACTD block of vehicles and ground equipment ordered by the Air Force. As a nonacquisition demonstration asset, GHMD could circumvent some of the exacting developmental and operational testing that would be imposed on a competitively acquired aircraft system. However, without the latitude for extensive maritime customization, the Navy chose not to deviate much from the basic Air Force Block 10 configuration. The Raytheon integrated sensor system came with an electro-optic/infrared and synthetic aperture radar payload. To that standard Block 10 package, the Navy ordered maritime modes installed for the radar that included inverted synthetic aperture radar.

The unique capability of inverted synthetic aperture radar so valued by Navy operators is its imaging of characteristics associated with ships' roll, pitch, and yaw by cancelling out background movement of surrounding wave action to yield a reasonably accurate classification, if not identification, of ship targets; an avionics set that combines the features of a radar warning receiver; an electronic support measures and electronic intelligence system; and an automatic identification system that tracks mandated emissions from ship traffic on position, course, and speed. The GHMD communications system was comprised of common data link for line-of-sight command and control and relay of imagery; Ku satellite communications link for up-and-away beyond-line-of-sight command and control and imagery relay; and Inmarsat satellite communications for backup command and control. The traditional Air Force launch and recovery method employed a line-of-sight command and control van, while the mission control element, housed in a separate mobile van, hosted command and control for beyond-line-of-sight mission monitoring and imagery download.

The Navy adapted the mission control element to its own tactical auxiliary ground station (TAGS) housed in a brick-and-mortar structure on base at the Naval Air Station Patuxent River:

> A tremendous amount of data is sorted at the TAGS: individual tracks are nominated to the Naval North Fleet East, which is Second Fleet's name when in its homeland defense role to U.S. Northern Command. Imagery is transmitted from the inverted synthetic aperture radar and electro-optic/infrared sensors to the Office of Naval Intelligence, Fleet Imagery Support Team and the National Maritime Intelligence Center in Suitland, Maryland, where imagery analysts examine the data and provide associations between the analyzed imagery and a target of interest. The two are paired and those tracks are nominated to a common operating picture provided to U.S. Fleet Forces Command, and disseminated over Global Command and Control System-Maritime (the operational architecture that provides worldwide connectivity for naval forces).[25]

To give GHMD a thorough workout, it was enrolled almost immediately after delivery to Patuxent River in an Air Force–sponsored exercise known as Joint Expeditionary Force Experiment 6 (JEFX-06). The overall exercise theme—networked warfighting—could not have been more appropriate to GHMD play. Its assigned role was to demonstrate maritime domain awareness as one element of the sensor fusion and dissemination of multiple sources of ISR information as another. On station, GHMD was required to "sort suspect vessels from all the rest and then have the ability to disseminate that information, not only among Navy stations and operations centers, but provide that information to inter-agencies," and in so doing to aid the Navy in "developing concepts of operation and tactics, techniques, and procedures" that offered promise of seamlessly transitioning to BAMS when it ultimately moved forward.[26]

Setting a vigorous pace, GHMD rolled from JEFX-06 to the multinational Rim of the Pacific exercise in August. GHMD captured images of a ship-sinking exercise, expanded maritime interdiction operations, and showed its wide-area search and surveillance capabilities to locate target vessels at sea. GHMD operated from Edwards Air Force Base in California to Hawaii for each mission, flying more than 2,500 miles each way for a total of more than one hundred flight hours and providing more than eight hours of on-station time during each mission.[27] The second GHMD vehicle, known as N2, arrived at Patuxent River in December 2006, marking completion of GHMD hardware delivery.[28] Throughout 2006, GHMD performance was viewed internally by Northrop Grumman as a lynchpin to a "keep it sold" strategy designed to consolidate BAMS discriminators when a full-up competition reemerged. Favorable contractor performance assessment reports—government evaluations that documented Northrop Grumman's management effectiveness in meeting required contract performance—could be a make-or-break determinant of past performance scoring in a future BAMS competition. Initial results were not promising.

GHMD struggled through start-up issues centered on reliability, availability, spares provisioning, and a host of problems that seeped into the GHMD program from ongoing difficulties with the Air Force Global Hawk program. Carl Johnson found himself under increased scrutiny by the Navy and his own management to bring the GHMD house into some order, requiring his near-full-time presence in Patuxent River for extended periods. Johnson persisted and, by the end of the year, the trend line was heading in a positive direction that ensured contractor performance assessment reports would not be unduly negatively affected.

By December 2006 significant and positive movement in the Navy's POM build for BAMS in FY2008 could be detected. Satisfied that the new manned maritime patrol aircraft had a secure foothold within the naval aviation budget, the Navy could now turn to resourcing the unmanned adjunct to the P-8—ironically, the adjunct that made possible the acquisition of the P-8 by lowering the required numbers and hence the budget outlay for an inventory of manned aircraft. Fiscal year 2007 funding levels for BAMS were carried over into FY2008 and were scoped toward what could be the start-up of competitive activity aimed at an eventual request for proposal solicitation to industry for system development and demonstration.

Significantly, FY2008 research and development funding was holding at $116 million—close to the prior year's amount, and indicative of anticipated growing contractor activity. Fiscal year 2009 funding almost doubled, adding $228 million—a sure sign of anticipated robust contractor activity in the latter phase of system development and demonstration, preparatory to a preliminary design review on the road to low-rate initial production.[29] Expectations ran high at Northrop Grumman with news of the Navy's new seriousness: the years-long interregnum—a thirty-six-month period of "wait-and-see" after the BAMS concept had been defined in concept study and affirmed in PUMAS, and a low point in which a succession of company officers failed to step forward and exhibit a strong leadership example that articulated a vision and strategy for BAMS—was now drawing to a close with the promise of renewed opportunity. Entering this more optimistic BAMS phase would prompt a wholesale changeout of key personnel on the company's BAMS capture team—a change already begun as a natural consequence of the lapse in expressed Navy interest. Most of the principal participants from Northrop Grumman and the Navy in the original BAMS study had retired or moved on to other projects. Ominously, the down time had allowed powerful industry competitors to mature alternative concepts. The competition would be a bellwether for Northrop Grumman: whether the grand vision for Global Hawk, begun as a preposterous win of the DARPA ACTD for tiny Ryan Aeronautical, could now translate to a dominant industry position in robust, large-scale operational UAVs by extending the product from one service customer to a second service customer.

CHAPTER 8
SHAPING THE CAPTURE STRATEGY

Though the Navy toiled for years to set up a competition for its Broad Area Maritime Surveillance unmanned aerial vehicle program, its long-awaited issuance of the request for proposals was without fanfare. The service quietly put the document on the Internet February 15.

—DAVID BOND[1]

At the middle of the first decade of the twenty-first century, most informed defense industry observers in Washington, DC, would not have prophesied a bright future for BAMS. Five years had passed since the original BAMS concept study had convulsed conventional thinking in the maritime patrol and reconnaissance community and injected an alternate future with unmanned aircraft into their thinking. Yet the BAMS initiative seemed to be losing ground, its prospects growing dimmer, as the Navy had in 2004 selected Boeing's 737 derivative, the P-8 Poseidon, as its manned aircraft for the future of maritime patrol and reconnaissance. The Navy had withdrawn its BAMS restart solicitation and seemingly had closed down further acquisition activity. The core group of disrupters in Northrop Grumman and in the Navy most responsible for establishing the BAMS concept in those earlier years—Frauenberger, Lapins, Snowden, and Easterling—had moved on to other positions, had retired, or were only maintaining peripheral association with the BAMS concept through work on the PUMAS study. If there were a clear path ahead for BAMS at this point, it was not evident to most.

However, a new cast member in this years-long drama was debuting, a disrupter in his own right who, when paired with Bob Mitchell, would play a preeminent role in the final act: the resurrected and reinvigorated BAMS competition. Bob Wood was already a familiar name in Bethpage as a business developer with enormous talent as a strategist and relationship builder.

Wood's flair for business development came to the attention of Northrop Grumman business developers as early as 1999. Frauenberger and Easterling were just beginning their dialogue on a UAV adjunct to the MMA manned aircraft as Wood was representing Boeing to the Navy in Washington, DC, for the F/A-18 Hornet program. For Wood, Boeing seemed the logical transition opportunity from active duty, during which he had commanded a Hornet squadron in combat and more recently was the Hornet requirements officer on the CNO's staff. To his growing marketing expertise, he could add established experience in the technical intricacies of aeronautical systems, the sometimes obtuse nature of defense acquisition rules, and an intimate knowledge of Navy organization and officer personalities. It was in his Boeing role that he came to the attention of Ernie Snowden at Northrop Grumman, due to their collaboration on Hornet strategy.

Among his pursuits outside of Northrop Grumman, Snowden was the head of the local chapter of the Association of Naval Aviation (ANA). It fell to him to organize, plan, and manage the group's annual national convention when it cycled to the Washington area in 1999. He had mapped out a fairly robust convention agenda under the organizing theme "Ready for the Next Millennium"—still too early in 1999 to appreciate the irony of a stalled BAMS program just five years later that belied that theme. Presentations and social gatherings were lined up to create encounters between senior Navy flag officers and many defense industry executives who were underwriting convention costs with corporate donations. But the agenda lacked an anchor event that would be the signature activity for this convention and a highlight for the 2,500 member attendees from around the country.

Snowden reached out to Bob Wood to ask his help to conceive and organize an event that might involve the Smithsonian Air and Space Museum. Wood sprang into action. He conceived the "Salute to Navy and Marine Corps Aces" as the event title, met with the museum director, retired Vice Adm. Don Engen, to obtain approval and recommendations, personally contacted more than thirty-five living aces and secured their cooperation and attendance, and commissioned an original artwork by renowned aviation painter Roy Grinnell that captured the essence of this celebratory evening. The aces all attended and personally signed each of several hundred prints that would become a moneymaker for the ANA chapter.

For the big evening, held in the IMAX theater at the Air and Space Museum, Admiral Engen presided and introduced a representative ace from each of the nation's major conflicts for short remarks, followed by an introduction of each ace in attendance as his early service photo in flight gear was shown on the IMAX screen—all accompanied by a patriotic medley by a local high school chorale at the back of the theater. The emotional impact and resonance of the event were unlike anything most attendees had ever experienced, and it was talked about for months by naval aviators and defense industry executives around the country. And importantly, it was a testament to Wood's flair for organizing momentous events that serviced a larger theme, a testament that was not lost on Snowden.

Four years later, when the E-2 program in Bethpage urgently required a seasoned business development leader, Snowden reached out again to Wood, this time to offer the position in Northrop Grumman. As the leader of E-2 business development, Wood had an enduring impact. Perhaps most memorable to the community were the annual supplier events that he created and orchestrated on Capitol Hill that brought in senior corporate officers from more than a dozen major subcontractors on the program to sit with members of Congress and senior Navy flag officers. The effect was electric and energized the supplier base to fan out and lobby their representatives on behalf of appropriations for the E-2 Advanced Hawkeye program. Wood's talents were becoming recognized throughout the integrated systems sector, and he would soon be called on again to lead a revitalized BAMS effort.

Nearing the close of 2006, the Northrop Grumman Corporation was still very much occupied with the absorption and rationalization of major new companies, including Litton, Newport News Shipbuilding, and TRW, acquired within in the previous five years. At the same time, Northrop Grumman's integrated systems sector was being buffeted by the loss of major anticipated new growth opportunities in its strategic plan with the cancellation of the Air Force's multisensor command and control aircraft, a Nunn-McCurdy breach (exceeding required cost thresholds) with its Air Force Global Hawk program, and a curtailed outlook for historic franchises in electronic attack aircraft and target drones. A startling realization took hold: the company had not won a major new unclassified aircraft competition since the award of the F/A-18 some thirty years before, when Northrop was the junior partner to prime aircraft manufacturer McDonnell Douglas.

As the company's strategic planners surveyed the outlook for new growth opportunities, the prospect of still leaner years of DOD budgets lay ahead; new opportunities for modernizing and recapitalizing the force were reduced to a smaller number of larger make-or-break competitive opportunities. To make the investment commitment required to be competitive for those opportunities required a careful consideration and prioritization of discretionary monies for marketing and selling, capital, bid and proposal, and associated independent research and development—though some of the latter could be recovered through the rate structure.

Recognizing this, the Northrop Grumman corporate leadership imposed a protocol for determining corporate must-wins—those opportunities that contributed substantially to the growth plan and cemented market share or expanded into adjacent markets. The result was three must-win pursuits for the integrated systems sector: the unmanned combat air system–demonstrator (UCAS-D), an unmanned aircraft that would be required to make aircraft carrier landings and takeoffs and was viewed internally as a logical expansion of the sector's market dominance in unmanned air system technology; KC-X, a large-body aerial refueling tanker aircraft for the Air Force, seen as the way to reenter large aircraft production using an essentially mature foreign aircraft entry represented by a modified Airbus A330; and BAMS, now reemerging in the Navy's acquisition planning.

In April 2007 Northrop Grumman submitted its bid for KC-X to the Air Force, and the Navy issued its final solicitation to industry for BAMS. Four months later, the first of the three sector must-wins was brought home with the award of the Navy's UCAS-D development contract. By then, the BAMS capture team was totally consumed in guiding their response to the solicitation through the postsubmission phase and was not wholly convinced that they had any kind of competitive edge. Notwithstanding the strong finish in the PUMAS analysis, noteworthy exercise results with Global Hawk Maritime Demonstrators, or the sector's expertise in unmanned systems, defense industry competitors were lining up to challenge each of those must-wins, including BAMS. The competitors, after all, were industry behemoths Lockheed Martin and Boeing, each with its own strong reputation, connections to Navy leadership, and deep pockets for investment to sharpen its discriminators. The essential ingredients for the Northrop Grumman BAMS capture team would prove, once again, to be strategy and leadership.

Since joining the Northrop Grumman integrated systems sector in September 2003 as the business development director for the E-2 program, Wood had seized on several opportunities to further mature his business development skillset and that of his team. They were executing a $1 billion program for the company annually, developing new technologies and new capabilities, all the while developing a new variant of the E-2C for the U.S. Navy—the E-2D Advanced Hawkeye. The program team was consumed with protecting its appropriated funds and requested budget for the Navy customer through the legislative process on Capitol Hill, attempting to sell more E-2s around the world with targeted customers such as the United Arab Emirates, India, and Malaysia, while also enhancing the aircraft's capabilities for existing international customers France, Egypt, and Singapore. After almost three years in the job and being involved in a number of smaller competitive captures and market campaigns for major subsystems and support for the E-2 program, Wood found himself in a great place with an excellent boss, good leadership team, amazing defense technologies, and a very important program for his new company and the U.S. Navy. Northrop Grumman had gone through almost a complete transformation in its business development processes and competitive capture training after suffering a series of important losses. As a result of a renewed emphasis on growing the business base and the more aggressive leadership style of Tom Vice, then vice president of business development for the integrated systems sector, business development instituted a series of capture training workshops, expanded the pool of business development staff with new hires, and stepped back to reflect on processes for how best to implement a competitive capture of government procurements.

An important ingredient for all new business capture initiatives going forward would be strict adherence to the logic and processes of an evolving but ever more comprehensive business acquisition process (BAP) that was being put into common usage throughout the sector. The hallmarks of this new focus were a greater emphasis on nailing down win themes and strategy at the front end and progressive evaluation and critique throughout by senior sector leader means using a system of successive "color" reviews. By the time of Wood's involvement in the BAMS capture, the BAP had evolved to a highly organized and structured handbook governing the actions of all who would be expected to have a role in the capture.

The process began with the formation of a pursuit team and approval of the pursuit, normally at the most senior level of the integrated systems sector, for opportunities determined to be in that sector's lane. Since BAMS had been conceived and nurtured by the integrated systems sector since the concept's genesis almost eight years before, it was a logical and straightforward determination that it would revert to that sector to establish the bid priority and marshal the resources now required to compete. With that acknowledgment and approval, and a competitive assessment pulled together by a competitor analysis group in sector business development, the capture plan and strategy, prepared by capture campaign lead Wood and capture executive Bob Mitchell, would get seriously scrutinized by an experienced sector senior staff in a blue team review—normally executive-level heads of functional departments—augmented by a gray team nonadvocate review by seasoned experts, most often the retired flag officer–level consultant review board. A series of well-delineated procedures then took the team through a sequence of preproposal actions and proposal development steps that required the standup of a pink team to ensure that the win strategy was well mapped to the proposal; a red team, usually a group of senior directors with deep acquisition experience, to emulate the customer's source selection evaluation board and verify the proposal's compliance; a nonadvocate review that focused on executability within projected cost; and a gold team review, again composed of executive-level functional heads or retired flag officers with acquisition experience that validated the final proposal product as meeting customer requirements.

Having been in several business development roles in his defense career since retiring from active duty in 1998, Wood now found the world of "business or strategic capture" an exciting domain in the aerospace and defense business. It blended all of the aspects of business that interested him: leadership, teams, competing and winning, customers, relationships, incredible technology products, and working to accomplish something very difficult. Of course, winning a competitive procurement is an extremely satisfying endeavor. Wood had watched other campaign and capture leads[2] at Northrop Grumman work on their respective efforts and over the years from his time at Boeing and Northrop Grumman had developed his own sense of the "art and science" of winning competitive captures.

IMPOSING A DISCIPLINED CAPTURE PROCESS

Periodically (generally quarterly), the leadership of the integrated systems sector reviewed ongoing business development and capture efforts.[3] These reviews involved the various functional leadership personnel of the businesses to include the general manager, business development, finance, and communications representatives, program managers, and so forth. The president of the sector generally oversaw these reviews at the business level, and this was where the focus on win strategy, probability of win, campaign strategy, customer contacts, and engagement was reviewed.

Also, at periodic intervals, a team of consultants called the board of visitors reviewed the capture efforts. This body was a holdover from prior years—acknowledged inside Northrop Grumman to have been modeled on a consultant review board put in place in the former Grumman unit in Bethpage for its effectiveness, but now empaneled at the higher sector level and with correspondingly more senior consultants at the corporate level as well. These were experienced senior leaders, either in business or retired from the military services. They brought direct customer knowledge to bear on the specific campaigns to assist with diving deeply into the strategy, tactics, and intricacies of the capture pursuit.

Both processes could be challenging for the capture leader, as it was generally a no-holds-barred leadership review of every aspect of the capture. No stone was left unturned, and the capture lead (either the capture executive or the business development campaign leader) landed in front of a difficult audience of experienced inquisitors. The process was a good one, for the capture team often would become so absorbed in the minutiae of everyday implementation that it would be difficult to stand back and see clearly where the strategic capture effort was going and whether the stated tactics were having the expected impact. On some occasions, the capture lead was not able to put the real work into the process of assessing the capture status and the details that had to be thought through to achieve success. Those occasions usually did not fare well for him or her and their team during the review process.

During one of these day-long series of campaign reviews, where Wood was briefing an E-2–specific capture effort, he watched the campaign leader at the time for BAMS, Lyn Whitmer (who was the sector representative to NAVAIR in the company's Patuxent River office), present a strategy and campaign review of the BAMS campaign in spring 2006. Wood thought at the time how painful the review was, and he was astounded by how far behind and how ill prepared for the impending capture and proposal process the company was. The all-out, make-or-break solicitation response was looming on the horizon. In fairness to Whitmer, he had been only recently drafted as the campaign leader for this important must-win, mostly on the basis of his program knowledge from his prior role on active duty as the Navy program manager. He was new to his civilian post, had not yet guided a competitive team on the industry side, and had not been given resources or regular access to senior sector leadership from his outlying post at Patuxent River. It seemed that the company was looking at a huge mountain to climb on this effort. The absence of a full-time marketing campaign leader since Ernie Snowden's departure to other duties in 2004 and the failure of UMS leadership in Rancho Bernardo to shore up the gap in the interim had put the effort well behind. The team was simply unprepared for a major proposal response and was not showing evidence of a plan to recover.

One aspect of the capture review that was disconcerting to most who had not been close to BAMS at the time was the impression that this proposal was Northrop Grumman's to lose and that the company should have been better positioned for the capture effort against the competition. Northrop Grumman had spent so much intellectual capital, time, money, and effort on creation of the BAMS concept, working with the Navy on the concept study effort and subsequent PUMAS analysis. How could the company not be in the lead on winning this extremely critical effort for Northrop Grumman—an effort that could redefine the business both financially and programmatically for years to come? Wood's thinking at the time was twofold: "How could we have let this ball fall to the ground so precariously since the start and stop of the program in 2004?" and, "Thank goodness I have nothing to do with this capture effort, for this is not going to go well!" This thinking would be seriously tested in multiple ways. Little did Wood know that he would soon be asked to take over and become the campaign lead for the BAMS capture.

On July 15, 2006, Wood arrived at the airport to check in for his flight to the Farnborough international airshow. Much work had been done to prepare for the airshow regarding the E-2 customer efforts and meetings. Wood and the team had an extensive schedule of meetings set up with international customers. The show setup was a massive undertaking, beginning the preceding weekend in preparation for opening day on Monday, July 17. Wood was involved full time over the weekend ensuring that displays, presentations, escorts, and briefers would be ready for Monday's agenda to kick off. As he proceeded to the ticket counter area, he received a call from his boss Tim Farrell, vice president and general manager of the E-2 program, who was already at the show attending some pre-show events. He informed Wood that a decision had been made to take him off of the E-2 program and have him take on the campaign lead role for the BAMS campaign. Wood was to get to the United Kingdom, and Farrell would discuss it with him at the show. Farrell already had some discussions with Tom Vice, the sector business development vice president, who convinced him that Wood was the right person to take on this effort. Lyn Whitmer, perhaps experiencing his own frustration at the apparent lack of corporate resolve to this point, had decided to leave the company and take a job in Lockheed Martin's Patuxent River office. This departure created an even greater urgency to close the gap for the BAMS capture effort, already behind in mobilizing resources for a successful capture.

Upon Wood's arrival at the company pavilion at the show, Vice and Farrell approached him to discuss the change in personnel and their reasons for it. He realized the magnitude of this decision on his part, given where the BAMS capture effort currently stood, the high visibility of this program, and the significance of the program to the company. With some trepidation he listened to their urging to take on this role, as he knew it would entail a massive amount of work based on his observation of the BAMS situation from the spring capture review. In the course of these discussions, Wood came to understand that the decision was not optional on his part and that he needed to accept if he was to further his career and maintain his stature in the company. Tom Vice was particularly not one to accept a refusal. He was extremely aggressive about demonstrating a record of accomplishment and building an upward trajectory to his career. He was clearly on a solid path to very senior roles at Northrop Grumman.

Wood left this conversation with Vice and Farrell feeling genuinely excited about the opportunity, which could only be interpreted as a vote of confidence in his abilities and reputation in the sector. But he had a gnawing anxiety over where and how to start with this effort, given its almost neglected condition and the impending proposal preparation. At the show, he spoke further with Vice on what the opportunity could mean to him professionally in the company. Vice assured him that if he took this on and won the capture, there would be future promotional opportunities. For the duration of the show, Wood continued in his role on the E-2 program. He was professionally satisfied and justifiably proud of his business development team's efforts in stabilizing the E-2 program and positioning it for solid transition into production and continued growth. His wife arrived on the last day of the show to join him for some vacation time in Venice, Vienna, and Munich. This relaxation was well deserved, but aspects of the job change dominated his thoughts during times of quiet reflection. Wood returned to the United States on July 24 to begin the turnover of his E-2 job to his deputy, and he started to spend some time in the newly assigned proposal center for BAMS in Bethpage.

The capture team being built from scratch from different parts of the company included several elements from the E-2 program, several from the unmanned systems division (who remained in Rancho Bernardo), and the core of the team membership from both locations collocated in the capture team spaces established in Bethpage. Most of the people from the E-2 program were more familiar with sole-source contracts based on the E-2 legacy with the Navy than with major acquisition category competitive procurements. The initial person placed in charge of the capture was a director-level manager from the E-2 engineering function by the name of Mark Kassner who was energetic, determined and worked extremely hard to ramp up the capture effort. Within several months after Wood took over the business development role and campaign lead for BAMS, the company put the vice president and general manager of the unmanned systems business line in the role as the capture executive to install much more experience for this critical effort.

A highly respected business leader, Bob Mitchell was the former chief executive officer (CEO) of Ryan Aeronautics. Wood had not met Mitchell before but had heard a lot about him and was excited that he would be a direct boss on the capture. They connected quickly early on, likely because of their previous military aviation experience—Mitchell in the Royal Air Force and Wood as a U.S. naval aviator. What became clear in early dialogue was that they shared a passion for competing and winning. The BAMS opportunity presented great potential for a big corporate win. The challenge stimulated that passion and provided the basis for a solid partnership to aggressively go after the BAMS program by putting together a well-thought-out strategy for bringing this home for the company.

Mitchell also had a significant background with Global Hawk, having originally guided the design of the air vehicle and its systems that was proposed to and selected by DARPA prior to Northrop Grumman's acquisition of Ryan Aeronautical. The Block 20 Global Hawk would be the air vehicle offering to the Navy, and Mitchell's experience on Global Hawk was invaluable. He also arrived with a background that was somewhat unconventional for an American aerospace and defense company executive: he had flown the British Lightning, a supersonic fighter aircraft designed for the Soviet threat in the Cold War. So they both had a fighter background, with Wood having experience in the F-14 and the F/A-18 Hornet. They hit it off immediately as the capture executive and business development campaign lead.

Upon his arrival, Wood reached out to key functional leaders inside the team and others outside the team with relevant competitive and customer intelligence to start nailing down the capture strategy. Even though BAMS had significant history in the company and many had worked on its development for some time, most of those original contributors had migrated elsewhere while the environment for BAMS had now evolved significantly. The capture lacked a thorough, updated strategy formulation based on the current situation with the NAVAIR customer. Wood set up an initial blue team strategy session with some team members and brought in consultants that had been recently hired to help in this effort. His efforts were aided immensely by the addition of Black Ram, an outside consulting firm that specialized in proposal preparation.

The proposal manager Black Ram assigned to the BAMS effort was an affable man with a wealth of experience in proposals, but he soon had a significant health issue requiring him to depart the team. Al Hutchins, a retired Navy captain, NAVAIR program manager, and consultant with significant capture strategy and proposal experience, had been supporting Wood on strategy development for the capture, and it was a natural step to ask him to be the proposal manager—a decision that would soon really pay off. A 1971 Naval Academy graduate and CH-57 pilot, Hutchins had retired as the program manager for the UAV program office and, from that very specialized experience, had a wealth of knowledge about acquisition, NAVAIR, source selections, and winning strategies, particularly as applied to the emerging area for the Navy of unmanned systems. He had a great mind for capture strategy but also for constructing winning proposals, having worked for Lockheed Martin on the joint strike fighter proposal out of Fort Worth, Texas. Hutchins knew how to design all the steps for a competitive win, had an extremely analytical mind, and had a technical background that was especially relevant to the BAMS competition. Like Wood, he also loved strategic capture. By this time, he also had consulted with a number of companies in the Washington, DC, area, served on their advisory boards, and brought with him the experience of having worked inside many corporate cultures to produce a string of competitive wins. Hutchins's contributions to the BAMS capture effort spanned multiple areas including win strategy, marketing, proposal development and editing, advising the executive capture team, and much more.

Early on, Wood and Mitchell had constructed a briefing for Northrop Grumman CEO Ron Sugar and his corporate professional council of senior executive direct reports. This briefing contained an initial technical approach with the Block 20 Global Hawk, leveraging the work already done on the earlier Block 10 Global Hawk in evaluation with the Navy as the GHMD using maritime sensors adapted for overwater surveillance. Wood recalled distinctly the ride with Mitchell to the Northrop Grumman corporate office in Los Angeles, during which they discussed their respective flying experiences. Mitchell was also a civilian pilot and owned several aerobatic aircraft, one of which was Russian made. Comparing notes on flying military aircraft, Mitchell asked Wood how high he had ever been in a fighter. This conversation made reference to the fact that Global Hawk flies at 60,000 feet. Wood answered that he had probably been no higher than 47,000 feet in his time flying the F/A-18 Hornet and the highest altitudes he had flown either tactically or on functional test flights. Mitchell allowed that he had been to 60,000 feet in a British Royal Air Force Lightning fighter. Wood was stunned by that fact. He had no understanding of the capabilities of a Lightning, but he was duly impressed, and this exchange further cemented their close relationship.

In the course of preparing for the briefing to the corporate leadership, Wood came to understand more of the philosophical underpinnings to Mitchell's approach to leading this major bid, an approach that could be summarized in just a few major tenets:

- The capture executive must be also the leader in strategic thinking—most of the decisions that rise to this level during the campaign are by nature hard ones that often alienate people.

- The win strategy must be compelling, with powerful, measurable, and defendable discriminators.

- Teaming decisions—whether externally or internally with other parts of the company—must be made on the basis of strength added to the team, not simply because someone wants a "piece of the action."

- The capture executive's tenure will be marked initially by reversing many earlier decisions.

- The capture executive will set the tempo from the start, and people should expect to be engaged around the clock.

- The capture executive works best with a strong adjutant and sounding board, preferably someone from business development who can bring an inside and outside perspective.

- The capture executive must expunge from the team any sense of entitlement and overconfidence in a win because of pride in their legacy product.

- While the corporate leadership can be an asset to be exploited, the capture executive must not be too controlled by the whims of corporate leaders.

The briefing at the corporate offices went well. Mitchell asked Wood to brief the campaign portion, which covered the win themes, messaging, customer contact objectives, shaping strategy, and details of the assessed competitor strengths and weaknesses and their likely strategies. The presentation that Mitchell and Wood gave that morning was persuasive in gaining unanimous endorsement of the win strategy. By then it was understood that the primary competitor appeared to be a Lockheed Martin and Global Atomics team offering a navalized Predator UAS that they had introduced as Mariner. Mariner was smaller and cheaper, potentially giving the Navy more air vehicles at lower unit cost. But their offering required some new development to make Mariner more effective in the maritime mission. The Northrop Grumman assessment was that Mariner was more limited in capability and could just make threshold performance in key areas with some modifications. The Northrop Grumman concept, by comparison, came with higher unit cost but could achieve objective-level performance with far less development cost to the Navy. And due to payload and power available, it had substantially more room for growth. As long as the Northrop Grumman offering could be kept under budget and within the Navy's funding profile and remain in a competitive cost range with Mariner, the strongest case for a win pointed to an evaluation based on best-value criteria. That was not a certainty at this point—many recent DOD major acquisition competitions appeared to be going to the lowest price bidder—so this became a primary focus of the customer shaping campaign as Wood portrayed it to the corporate leadership.

To win on a best-value approach, Mitchell, Wood, and the capture team had to show that their technical offering would be the optimal choice for the Navy's maritime patrol and reconnaissance mission. This required that key conditions be set early, and these became part of the detail that the corporate leadership gained insight into and approved: the team would take a nondevelopmental low-risk approach that required minimum change to the baseline RQ-4B aircraft system; by the proposal submission, they would maintain or achieve technology readiness level six or better across the board, thus reducing perceived risk; they would meet all threshold requirements and more than 90 percent of objective requirements; they would build sensor configuration around a 360-degree active electronically scanned array radar that met threshold, objective, and growth requirements; they would corporately invest in a set of highly visible risk-reducing tasks to remove all potential red or yellow risk items; and they would build a full-scale developmental test article, dubbed white tail, to tangibly demonstrate the marinization modifications. Programmatically, they committed themselves to accelerate the Navy's declared initial operational capability date by one year, from FY2013 to FY2012.

Finally, the team made the location for BAMS activity once under contract—especially production and final assembly—an element of the strategy by offering to site the program in a locale that offered the greatest benefit to the Navy for cost, technical, and political factors. There was a lot of discussion around win strategy, competitive discriminators, the government timetable, the competition's strengths and weaknesses, the campaign contact plan, and the two Block 10 Global Hawks that the Navy had stationed at Patuxent River for use as demonstrators to inform operating concepts for BAMS. It was clear that Wood and Mitchell now had the entire corporate leadership behind them and that this was a significant must-win for the company.

The proposal team was set up in an aging building on the Bethpage campus, a place much diminished since Grumman's heyday some thirty years before. The team occupied an office space in building 25, the same building, ironically, that Grumman had devoted to design and final assembly of the Apollo program's lunar module. During the standup, the proposal team began to put more work and detailed thought into the strategy. The team grew quickly as Mitchell brought in some of his former Ryan Aeronautics people and colleagues involved in Global Hawk engineering, logistics, and sensors. Since this was going to be a highly technical proposal that also included a classified volume, immediate standup of the full team was imperative. A draft request for proposal solicitation to industry had already been issued by NAVAIR. There was much to do, and the team was way behind the proper timetable to shape the procurement with the Navy and start to write a draft proposal. The objectives that needed to be quickly achieved almost seemed impossible given the time constraints and the looming release of the final RFP by NAVAIR.

As the capture team assembled, Wood began to observe the unique and disparate levels of experience and background among new members. As he met a number of engineers and technical people from the West Coast unit of the sector, it became clear that a significant rift existed between the UMS people in Rancho Bernardo and the airborne early warning/electronic warfare division staff in Bethpage. He was quickly discovering that relations between the East Coast and West Coast employees had been strained and rancorous since the earliest days of the BAMS concept study, owing in part to lingering ill feelings in Bethpage for the manner in which Northrop had subjugated them in their acquisition.

In addition, during several early discussions on proposal themes, the former Ryan employees seemed overly prideful, even arrogant, about their Global Hawk platform and technologies in the unmanned systems product line. There had been contentious discussions around who would lead this capture for the Navy's BAMS procurement and who really had the most to offer from a technical, win strategy, customer relationship, maritime surveillance experience, unmanned aircraft systems background perspective. Scott Seymour, the sector president, ultimately decided that Bethpage would lead it, but a real level of frustration and anger over this decision from the UMS side (mostly former Ryan Aeronautics people) simmered below the surface. This rancor would later come out in the tension and stress around writing the proposal, erupting as vocal arguments and heated confrontations in the proposal center that mostly centered on how far the team would go in compromising the Global Hawk design toward meeting Navy requirements. It would be a constant source of conflict, an unnecessary distraction for Wood and Mitchell in their efforts to build a cohesive team that was single-minded in its dedication to winning this competition. Here, the absence of clear direction from leadership at the sector and corporate levels to set aside any turf claims and to collaborate fully for the best sector outcome was sorely felt. The disruption and incompetence evident in the

Northrop takeover of Grumman more than a decade before apparently had never been settled, and it persisted as several generations of senior Northrop Grumman leaders turned a blind eye to the fault lines that still existed.

After the blue team strategy review, the amount of work ahead of the team seemed insurmountable in some ways. The strategy was immature and not concise and clear enough to establish a framework for building out a coherent set of messages that withstood customer scrutiny and the day-to-day tactics that would allow the team to socialize key messages. The team had not really engaged with the customer in the recent past—not since before defining those discrete discriminators that could be articulated in easily understood and welcomed themes for Navy customers.

Importantly, these themes would also guide proposal writers in preparing for more exacting critique in the actual proposal evaluation process. There was already a draft RFP on the street, and the team was composing a draft proposal in response despite not yet having a real team built and with no thematic guidance to proposal volume writers about how to organize their narratives around win themes. The technical changes needed to modify the Global Hawk air vehicle for the BAMS offering had not been thought through. That and other complex engineering issues that were still unsettled had to be worked through to meet compliance with Navy requirements. Meetings with key NAVAIR and CNO's staff requirements people were not happening and had not been a priority for the preceding few years. Turnover in the relevant Navy offices in the meantime meant that many Navy individuals who would likely be influencers in the competitive evaluation were new to a Global Hawk–centered BAMS concept. They had not yet been contacted for a briefing or concept overview. Nor had there been a concerted marketing effort that disseminated media products to reinforce the team image. Wood understood immediately the critical need for engaging his core business development group and the sector communications people to focus on that shortfall. The team needed to employ some unique tactics to catapult them back into a highly visible position in the competitive field. First and foremost was securing appointments with the NAVAIR leadership and program office.

The program manager, Capt. Paul Morgan, a former P-3 aviator, was known to be rather challenging with regard to his leadership style within the endurance UAV program office. As a first order of business, Wood and Mitchell scheduled a meeting with him at the NAVAIR headquarters building on Naval Air Station Patuxent River. As they were walked through the program spaces to reach Captain Morgan's office, they noticed several prominent signs: "We are in a BAMS lockdown. . . . be careful what you say in open spaces. . . . There could be contractors in the office spaces." This was Mitchell's and Wood's first evidence of the government's concern that there was a significant chance of a contractor protest. The program office seemed to be overreacting to the possibility that even dialogue on their requirements increased this likelihood. The Navy and NAVAIR did not like protests of competitive awards and were determined to do whatever they could to prevent this from happening.

After Captain Morgan greeted Mitchell and Wood cordially, they got down to discussing some of the various initiatives planned in the runup to the competition. They used the occasion to communicate what they felt were some of the strengths of a Northrop Grumman Global Hawk BAMS offering. As part of the discussion, they mentioned some of the senior customer meetings needed to gain a better understanding of the Navy's requirements for BAMS. Among those was a desired meeting with Rear Adm. Brian Prindle, then head of the Navy's maritime patrol community and the senior leader responsible for the fleet's position on BAMS requirements. Upon Wood mentioning this request, Captain Morgan flatly informed them that they would not be able to meet with Rear Admiral Prindle, as the draft RFP was out and there was now a lockdown on information to contractors. He added that the Navy was moving into the competition phase, and no contractor could be perceived as having an advantage over another.

Wood was incredulous over such a restriction. During the initial phase of a formal competition, this was the period of time during a draft RFP that contractors should be exploring solutions with their customers so that they can get a clear understanding of what the customer wants as a total system solution.[4] Wood said to Captain Morgan, "I am afraid, Captain, that such a restriction will not work and we will absolutely have to meet with Rear Admiral Prindle." Morgan appeared astonished that a contractor was defying his command not to meet with the admiral, and Mitchell looked surprised that Wood would challenge a key customer in this way (he noted later that he completely agreed with Wood regarding the position he took with Morgan). However, because the company had been so late in putting together a renewed capture effort that was comprehensive in pursuing every aspect of winning this procurement by shaping customer expectations, they did not immediately begin meeting with many of the customers involved in source selection. They believed, because of coming from behind, they were sure to lose this competition—a notion they could not bring themselves to contemplate.

This was a defining moment for Wood and Mitchell with the principal customer involved in source selection, and if they did not stand their ground at this juncture, it would be even more difficult down the road. Wood told Captain Morgan that because Northrop Grumman was late in pulling together their proposal team, it would actually be unfair to the company with regard to the other competitors if they were not permitted to meet key Navy and Pentagon leaders. They knew (as did Captain Morgan) that both the Lockheed Martin/General Atomics team and Boeing had already been meeting with many of the leaders they were intent on seeing, including Rear Admiral Prindle.

Captain Morgan thought about this and then agreed that they could meet with Rear Admiral Prindle, with the stipulation that they not discuss BAMS directly and only discuss the ongoing GHMD program and its associated aircraft now based at Patuxent River. There was great interest in the program office and in the Navy more generally about how these aircraft and associated ground control systems were performing, and especially their applicability to overall maritime surveillance. Northrop Grumman's role as the GHMD prime contractor was just an accepted fact of this procurement, and no competitor could dispute this program of record and its role in the ultimate development of requirements. They agreed to Morgan's demand but knew that this overly onerous and likely unenforceable restriction could be obeyed as requested, but would likely be skirted in the execution. Once they started talking to customers about GHMD, the eventual discussion would likely lead to BAMS. And this would open the door to explaining capabilities and differences that would be proposed for the BAMS solution.

As Mitchell and Wood began to execute the customer contact plan, they continually refined it to produce an increasingly robust and comprehensive list of both requirements and acquisition leaders, uniformed and civilian. These customer contacts represented leaders involved in some way with the BAMS procurement, either directly or indirectly, but also included many who in some way could be influencers of the leaders involved in deciding the best offering for the Navy and could exert sway over the decision process. In a highly competitive major acquisition category procurement such as BAMS, this was a common campaign practice, critical to the ultimate win by the primes and their teams by getting their marketing messages out. Influencers would often be other senior military service and Pentagon officials, congressional members and state and local officials whose districts stood to benefit economically, and an array of stakeholders who were identified as having something to gain or lose with a Northrop Grumman win. For BAMS, these highly placed Pentagon influencers were typically the assistant secretary of defense for acquisition technology and logistics and the assistant secretary of the Navy for research, development, and acquisition. Even the assistant secretary of the Air Force for acquisition, technology, and logistics could be an influencer, given the Air Force's potential economic benefit from economies of production and logistics support if the Navy bought the Global Hawk air vehicle. For congressional constituent interest, the Northrop Grumman team had to examine every supplier involved with manufacturing the entire Global Hawk air vehicle and associated ground control systems. This included airframe locations in Mississippi, California, and West Virginia. Industry primes and their subcontractors had always pursued the practice of shaping the customer toward their unique offering by highlighting the economic benefit of an award to businesses in their districts. There were different schools of thought as to the effectiveness of this practice, but experience shows that if the relationships have been established, protecting funding appropriations once the competition is won is easier.

Clearly, if success in business is a result of close personal relationships, there is more potential for success in closing a deal if closer ties have been established with one's counterpart in the negotiation. If people do business with people they like, respect, and admire—at times regardless of the real advantages of the technical approach or offering—then there is an incentive to pursue that campaign effort with vigor. This is exactly what

When we examine why companies win or lose new business in the aerospace and defense market, the reasons are very similar. Companies win more often when they focus on relationship, understanding the customer, and comprehending and responding to the customer's requirements. They lose more often when they don't execute these critical tasks. Similarly, qualifying new business opportunities early in the business development life cycle results in higher win rates, while late qualifications result in fewer wins and cost increases in business development. These and other activities are strong indicators of how well a company will do in competitive procurements. This correlation provides clear evidence that companies can raise their P_{win} by performing certain activities well and in the right sequence, thereby establishing the basis for an efficient process in capture management. Each company implements its capture management process to fit its own system, culture and management structure, and every implementation should include the same basic activities. So, Northrop Grumman and any company should be able to implement a strategic capture process that fits objectives and growth goals. The capture and business winning process is important, but it is not the "be all and end all" that guarantees a win. It provides a framework and discipline for a winning spirit and higher percentage of success. For BAMS capture, the Northrop Grumman team almost forgot all of this and came dangerously close to dropping the ball.

—ROBERT F. WOOD, "Reflections of a Campaign Leader," 2008

the team did with a vengeance. What really drove them this way was not only their overall campaign strategy but also the fact that they believed their competitors were ahead in doing this very thing and that Northrop Grumman needed to catch up. Some of their consultants and advocates in Navy offices with which they were already familiar implied that competitors were gaining on them or were ahead in this regard. Their counsel was that the team was losing this competition and that the competitors were closer than Northrop Grumman not just to the direct customer at NAVAIR, but also to the CNO's staff for aircraft requirements in the Pentagon. A collective sense of paranoia began to develop internally on the Northrop Grumman campaign: they thought they were losing the competition. The feeling was probably legitimate in that their competitors were prime industry giants on major naval aviation programs, and the relationships built from those programs were longstanding and personal, with Navy staffers in positions to overhear insider discussion.

CHAPTER 9
COMPETING TO WIN

If you want to escape the gravitational pull of the past, you have to be willing to challenge your own orthodoxies, to regenerate your core strategies and rethink your most fundamental assumptions about how you are going to compete.

—C. K. PRALAHAD[1]

The development of the Northrop Grumman BAMS win strategy, win themes, marketing and communications strategy, and the value proposition—what the team believed were table stakes, or customer desires that likely would not be compromised—began to be thoroughly and comprehensively formed in the blue team review, done in conformance with the sector's business acquisition process. Some initial work had been accomplished before the BAMS capture team was formally anointed and coalesced around an anticipated RFP. Now that a draft RFP was on the street, the team's work began in earnest.

A large aerospace and defense company working on a competitive procurement such as BAMS will very often invest immense intellectual capital, time, and resources on winning the opportunity. The results of the blue team were endorsed by the sector leadership and the strategy was set and promulgated, but numerous unforeseen obstacles appeared during the execution of the strategy. These obstacles could be daunting and often required a change to the assumptions that drove the strategy. Each day as the team worked with the Navy, encountered the competition in Navy spaces, received an update from the Navy contracting officer on some aspect of the draft and final RFPs, or heard some rumor from a field office or from a media outlet, they were swept by a roller-coaster of feelings of elation that they were winning and despair that they were losing. Mitchell brought with him a view that strategic capture of government business was like combat, whereby you are faced with executing a strategy and making the right winning moves only to be confronted by changes in the environment (the customer) or a new move by the enemy (the competitors) that required the team to adapt, refresh the strategy, and defeat or circumvent the new challenge. This was the approach of the entire capture team execution on BAMS—a long war with successive battles accompanied by short-term wins and setbacks to be dealt with.

From Wood's perspective, it was the team's strategic approach of taking the long view, all the while mindful of that long war, that ultimately led to success in winning BAMS. It clearly rested on precisely defining the desired outcome, properly framing the must-dos to achieve that outcome, and never deviating from holding that outcome fixed. This was a must-win for the BAMS capture team and for Northrop Grumman, and this was the only way to tackle it and win. The unknowns at the beginning were how long it would take and when victory could be declared, but with a leadership-endorsed and -guided win strategy and step-by-step execution mapped out for the team to work to, it would be achievable.

DIFFERENTIATING WITH PROVEN DISCRIMINATORS

In preparation for the blue team review and during the review itself, the team believed that their strategy needed to be simple, clear, and actionable. It needed to revolve around a value proposition with the NAVAIR customer that recognized and accounted for its table stakes. Win strategies are sometimes either too detailed or too vague, involving things that the capture team will do rather than actionable items that drive win strategies. For this team, it obviously centered on the Global Hawk air vehicle, which had already shown itself in Air Force operations to be the preeminent unmanned aircraft system ever designed for high-altitude ISR. The very capabilities of this platform were assessed internally as discriminators for Northrop Grumman. It was fast, powered by a muscular Rolls-Royce AE2000 turbofan engine, and climbed quickly to high altitude. Its endurance was impressive, having set a world record for airborne flight time of thirty-two hours without refueling. Besides its speed to station, it was a large air vehicle with a 131-foot wingspan and tremendous size, weight, and power-cooling (SWaP-C) that accommodated significant growth capability in avionics payload and for additional stores if needed. These were real discriminators in the view of the team, as one of the Navy's threshold requirements was the ETOS metric.

Critical to that requirement was getting to station expeditiously for one of the five orbits around the globe. Competitors Lockheed Martin/ General Atomics and Boeing were building or proposing air vehicles that were less capable than Global Hawk in these areas. Mitchell and Wood knew from competitive intelligence that Lockheed Martin/General Atomics would propose a maritime variant of their well-known and highly publicized Predator unmanned aircraft, which they had named Mariner for this Navy application. Mariner had an excellent time on station and endurance performance, but as a turboprop, it was much slower, and it was incapable of operating at Global Hawk altitudes of up to 60,000 feet. The Northrop Grumman team assessed that Mariner was going to be challenged in meeting an ETOS threshold requirement of 85 percent, a Navy standard at this point in the capture. Boeing was offering a G550 business jet manufactured by Gulfstream, configured as an optionally manned or unmanned version and militarized with unique sensors. This was a unique, even unconventional, offering that, given the Navy's requirement for an unmanned system, initially drew some derision from observers of the BAMS competition. But the team appreciated that Boeing could be a formidable competitor and very adept at shaping the customer's perceptions of their desired capabilities.

What was initially unknown was Boeing's sensor offering, which could also be a key discriminator. Boeing, always very secretive in its strategic captures, joined this competition late in the game, complicating the ability to acquire reliable intelligence on them and assess their offering. But it was soon apparent that their air vehicle would be better able to meet the ETOS requirement than the General Atomics Mariner system. Still, neither offering could match Global Hawk's ability to meet two key variables of the customer's value proposition: speed-to-station and room for growth.

As the campaign leader responsible for the "up and out" part of the strategic capture, including implementation of the capture strategy and customer engagement, Wood's attention turned to hiring someone with a business development background to support him in the strategy execution and the marketing and communication plan. His first offer went to a retired Navy captain, a former EA-6B Prowler pilot and ship commanding officer. With his deep naval aviation experience, Wood believed this person could hit the ground running in terms of quickly establishing relationships with the broader Navy unmanned aircraft systems community. And he was already in the company working in a branch of the same sector in Melbourne, Florida. Wood knew this individual from contact during their Navy service, and he managed to work out an arrangement whereby the newly hired individual would temporarily commute to Bethpage to work on BAMS for six months to help reach a more mature proposal. This arrangement worked only briefly, as the individual was more intent on retiring completely in Florida. This episode reflects one part of the complexity of running a large competitive capture: bringing on board and retaining the specialized talent that permits the team to run smoothly and cohesively to the finish line. Now, Wood had to redirect his attention to filling the personnel gap. He turned to an individual already with the company as a business development manager in the West Coast UMS division, Tom Twomey.

Twomey had prior Navy experience as an F-14 radar intercept officer but, importantly, also had interfaced for Northrop Grumman on Global Hawk with the Air Force customer and on GHMD with the Navy customer. Twomey brought great customer experience, product knowledge, and real energy and enthusiasm to supporting team efforts on the strategic capture and marketing communications activity, and he stayed with the capture all the way to proposal submittal. Without Twomey's help and experience, particularly his standing with the unmanned aircraft systems division in Rancho Bernardo, Wood could have anticipated more challenges in executing the strategy.

The blue team review was the gate that, once passed, would really set the capture activity in motion. Emerging from the blue team review, Wood briefed several consultant review boards, leading to many hours of discussion with the consultants to factor in their ideas and recommendations. The win strategy began to mature to the point where it had been distilled to one slide. And it was actionable, meaning there were clearly defined activities with expected outcomes that could be tracked in terms of assigned responsibility, resources required, and effects measured. The key aspects were price, technical factors, performance, and design of an early program to showcase capability to the Navy. This became known colloquially inside the sector as the Head Start program. Mitchell had used this strategy previously on other programs to separate out perceived high-risk items and then prioritize investment in risk reduction steps that could be used to ease customer concerns. Head Start also had the intended additional benefit of serving as a means to engage the customer and inform him of the team's intent and commitment.

Winning on price involved determining the Navy's sensitivity to cost and available budget and setting a price that was within the Navy funding profile and competitive when compared to the other offerings. Wood and Mitchell did not believe their offering would be the lowest price, but they wanted to be close enough, at least when pricing the system development and demonstration phase, low-rate initial production phase, and total operating costs in which they felt Northrop Grumman might have an advantage over the competitors. Everything depended on how the Navy computed its available budget, its capability priorities and tradeoffs, and the Northrop Grumman price offering. This became a constant source of the team's shaping efforts: to understand where the Navy's analysts were likely to land on this. Since the evaluation criterion was clearly stated to be best value, not lowest price, Mitchell and Wood felt the margin of capability superiority provided by Global Hawk BAMS in the maritime mission was going to be a winning advantage. If airplanes are bought by the pound, which is what the military essentially does, then the larger air vehicle is usually at a disadvantage on cost. Mitchell and Wood always believed that the Lockheed Martin/General Atomics Mariner, being smaller in physical size and weight and powered by a more economical turboprop, would be cheaper when considered as a basic air vehicle offering.

For the Northrop Grumman technical approach, the team adopted a nondevelopmental low-risk approach: the Block 20 Global Hawk, then the newest block configuration in initial production for the Air Force. An objective outlined in reviews for the team was to accept only minimum change to the air vehicle. It had to be at technology readiness level six maturity or better across the board, including sensors and systems. In addition, the win strategy required that the offer to the Navy meet all threshold requirements and 90 percent of objective requirements. By exceeding threshold and meeting most of the objective requirements and, further, by demonstrating reduced risk through the Head Start initiative, the team felt they could make themselves highly competitive in the technical space.

With regard to performance of the payload sensor system, the radar was decidedly a key element. It became a constant source of consternation and debate among the team and sector leadership. The current fielded state-of-the-art radar was a mechanically scanning radar, much like the conventional legacy fighter radars of the era. The debate here centered again on price. Mitchell and Wood were very concerned that they were not competitive on price, and the cost of the sensors was a big part of that. But they also wanted to be clearly ahead of the competitors with proposed surface search radar technology to make it compelling and attractive for the Navy customer. The Northrop Grumman electronic systems sector had been working on a funded project to develop with the Naval Research Lab a 360-degree active electronically scanned array (AESA) radar that was state of the art and could meet the technology readiness level requirements for the bid. It had flown on an airborne test bed several times, and the electronic systems sector had been working diligently on the technology. For an overwater maritime search capability, an AESA radar that could scan a 360-degree field of view was considered leading edge and did fit the Navy's objective requirement for BAMS. Credit goes largely to Mitchell for making an audaciously gritty strategic move to adopt this technology for its perceived value to this customer, even knowing that it raised the potential risk rating that the offering would draw from the evaluators and would also increase the price.

The choice was contentious, and a contingent of middle management on the team wanted to fall back to a cheaper mechanically scanned array that was off the shelf and fit under the assessed risk profile. The thinking was that once the competition was won, negotiations could be undertaken to fund the 360-degree AESA down the road in a spiral development. But Mitchell had convinced corporate and sector leadership that this was the way to get to the sweet spot of mission capability for the maritime patrol and reconnaissance community and, ultimately, to win.

The final decision did not happen overnight. It involved many capture team and corporate leadership meetings. Al Hutchins, the proposal manager, was also heavily involved in these leadership meetings and, given his technical acumen, ended up being a strong proponent of this approach. The resolution occurred after a senior-level meeting between integrated systems president Scott Seymour, electronic systems president Jim Pitts, Mitchell, and a select few senior consultants. Pitts was adamant that the 360-degree AESA be offered in the proposal as it was a critical technology for his sector's strategic growth plan and could be an incredible revenue generator, with future sales expanding on the back of BAMS. The word through the proposal center following the meeting was that it was contentious to the point that Pitts said he might offer the AESA radar to a competitor if the integrated systems sector did not use it on its Global Hawk BAMS offering. The electronic systems sector traditionally operated as a merchant supplier and had no qualms in going against another Northrop Grumman sector in a given competition if they negotiated a place as a supplier on a competing team. So the decision was made at the corporate CEO level to go with the 360-degree AESA.

A new name was coined to attach to the BAMS AESA radar that was picked up by the Navy customer in their references to the system: the multifunction active sensor (MFAS). The capture team was all in and went to work on the technical approach. The adoption of MFAS as the preferred system in the BAMS offering gave the team a lift and became a centerpiece of the Global Hawk BAMS concept, without which their preeminent technical standing would have become a more difficult case to make. Much of the discussion with the source selection team at NAVAIR following the award to Northrop Grumman centered around the fact that the MFAS radar was the discriminator that made an immense contribution to winning the program. The radar selection is a good example of how companies and capture teams struggle with configuration decisions of this magnitude on a platform bid such as BAMS.

During this same time, Wood became aware of a significant concern within the program office but reflected as well in the air vehicle structures competency of NAVAIR: that serious issues with Global Hawk air vehicle fatigue life and structural loads operating in the low-altitude environment had developed. The Lockheed Martin/General Atomics team had been talking to the customer about the need for the UAVs to descend to low altitude over water to obtain a visual identification of a contact such as a ship or submarine. Higher altitudes could present challenges with the onboard sensors, particularly electro-optic/infrared cameras, if they had to peer through inclement weather or a solid overcast. This prevented seeing the contact visually after initial detection by the maritime radar. The Lockheed Martin/General Atomics team had convinced the Navy customer that this should be a hard and fast requirement, which its Mariner offering could easily meet.

As the team watched this requirement solidify through the requirements process, ultimately documented in the draft RFP and systems requirement document, they became very intent on countering any negative information about Global Hawk's ability in this regard. An old internal Northrop document written by one of the original engineers was obtained that described possible problems with Global Hawk fatigue life in the low-altitude regime over water. There, turbulence could be significant with thermals and atmospheric disturbances over the world's oceans. Somehow, this memo made its way into the hands of NAVAIR aerostructures experts, who were starting to evaluate it. This serious issue for the Northrop Grumman BAMS capture team had to be addressed immediately before it got out of hand and became a disqualifier in the technical evaluation; doing so was critical in events that could completely derail a major competitive capture effort. The team could not imagine how this dated memorandum got into the hands of the Navy officials involved with the BAMS procurement. It generated a lot of discussion and stress. Given the experience the capture leadership team had on the design and development of Global Hawk, they knew they could counter this incorrect information, which was actually an internal engineering memo written around a recommendation for further analysis.

As the campaign lead, Wood knew there was a way to get in to see the aerostructures group. From Wood's earlier work on the F/A-18 as a Boeing employee, he knew personally Rich Gilpin, former Navy civilian working on the Hornet and now a member of the senior executive service in charge of the aerostructures branch. He contacted Gilpin and said that his team needed to come down with their engineers to brief him and his team on this issue. Wood wanted Gilpin to be completely aware that the original memorandum that sparked this concern was not an accurate engineering analysis of this issue and that Global Hawk did not have problems with either load factors or fatigue life over water. Gilpin agreed to see the team but was clearly nervous about it, considering the clampdown on communications with industry during this phase of the competition. Wood brought Alfredo Ramirez and several of the Ryan engineers to brief the structures group and Gilpin's immediate staff. After several hours preparing and rehearsing the briefing, to include consideration of all possible queries from Gilpin's engineers, the team traveled to Patuxent River to spend several hours with their NAVAIR counterparts.

The meeting went well; the team was able to dispel poorly developed facts and debunk the innuendos about Global Hawk's performance and structural integrity. In fact, it was so successful that it laid the foundation for the ultimate proposal narrative on Global Hawk's air vehicle structure, a narrative that was surely evaluated by the NAVAIR structures engineers. Even though this was an unmanned aircraft, and a pilot who might be at risk due to inadequate structural design was never intended to be on board, concern remained about a midair mishap and possible damage or injury to personnel on the ground caused by mechanical failure. In hindsight, this event seemed like a minor obstacle for the capture team, but in real time it generated a lot of concern. It also highlighted the close bond of the team and how it pulled together to resolve a crisis—one of many it would encounter over the next two years.

The Head Start initiative relied on corporate investment, which meshed well with an internally funded independent research and development project. It was used to demonstrate to the Navy customer Northrop Grumman's sincere commitment to lowering risk by conducting airborne demonstrations of preferred sensor and communications technologies on a corporate-owned Gulfstream G5 business jet. The initiative was briefed internally for approval at the sector and corporate levels with the understanding that it was going to be expensive. But everyone involved in the reviews understood that this kind of extraordinary investment was necessary to separate the BAMS offering from the competitors' offerings, to underscore the commitment to supporting a Navy mission, and to ensure better-than-even odds of favorable scoring in the proposal evaluation. The intent of Head Start was to resolve all red and yellow risk items identified with the technical solution and then to find a way to characterize the reduced risk and cost avoidance for the government in the proposal to the Navy. The team was not interested in publicizing this effort before it was completed, figuring that the competitors would find out and go to work shaping customer expectations by refuting Head Start claims or by inventing their own similar initiative. To preserve confidentiality on Head Start, the aircraft was flown out of a secure facility in Palmdale, California.

The critical need was to showcase this effort to the Navy to highlight a discriminator. Part of the Navy vision for the BAMS system was connectivity to a ground station for exploitation of ISR data from BAMS aircraft sensors. Doing so in a live demonstration of connectivity from an airborne surrogate could be used to reinforce the assertion that Northrop Grumman, more than the competitors, understood the BAMS mission. The Head Start project was put under the oversight of one of the capture team engineers. It proved to be a brilliant strategic move and ultimately paid off, despite the cost and difficulties of executing it. The plan was to do an experiment whereby the testbed aircraft would do a data link connection from flight to a ground station on the East Coast. This effort became a substantial part of the strategy, yet it was fraught with risk and required significant sector investment and the accompanying scrutiny.

Mitchell and Wood were not sure it could be pulled off, particularly with the public nature of the demonstration using the BAMS test bed. The plan was to conduct a public experiment in front of journalists based in the company-built BAMS ground station or mission control segment positioned in the parking lot of a local subsidiary office in Patuxent River, Maryland. This subsidiary, PRB Associates, had a unique history of supporting the Navy's maritime patrol community with work on tactical control stations and mobile operations command and control centers. This subsidiary office had been acquired within the past several years, and its specialization in technologies in the ground station domain made it the logical host for the BAMS control station mockup and a perfect tie-in to a subset of the Navy maritime patrol and reconnaissance community.

The Head Start investment involved not only the modification of the Gulfstream G5 aircraft, but also the construction of a domed ground station full of displays and communications technologies designed to showcase the team's understanding of maritime ISR in the eventual offering for the Navy. This ground station soon took on the name "podule." It was common understanding that the Navy customer was in communications lockdown imposed by the acquisition rules and that they would not set foot in the podule to observe this demonstration. But Wood conceived two other means of highlighting the demo in a way that would be seen by the Navy: inviting a well-read and highly regarded journalist to attend the event to provide press coverage, and ultimately finding some way to write in detail about this in the proposal to the Navy. Both plans were eventually successful.

Aviation Week & Space Technology magazine journalists were offered exclusive coverage of this demonstration if they agreed to highlight it as a major piece in their weekly magazine, one of the premier aerospace journals around the world read by nearly everyone in the aerospace and defense industry, including the Navy customers. The journal accepted, and Wood's communications staff worked hard to obtain a commitment from *Aviation Week* journalist Amy Butler to visit the podule and to monitor the event from inside that mission control segment in the parking lot at Patuxent River. Simultaneously, Guy Norris, another *Aviation Week* journalist, was invited to observe the demonstration from inside the Gulfstream G5 airborne testbed during the demo flight. So on a beautiful summer day in Palmdale at the Northrop Grumman facility, the completely modified G5 took to the skies over California, equipped with both the MFAS radar demonstrator and communications architecture intended as the baseline system for the BAMS offering.

There were some initial communications challenges with the G5, and as Amy Butler took her seat in the podule, Wood was reminded of the risk of inviting the media or customers to a live demonstration; there is always a chance something could go wrong. But a communications link with the G5 was finally established, and radar images and data were transmitted from across the country directly to the ground station in full view of Butler and, through her words, her *Aviation Week* readers a few weeks later. This huge success was a high point in the capture effort—the shaping of the customer, potentially in a very big way. As near as could be discerned from competitor intelligence inputs, no other competitor had gone to this length or invested like this to ensure that the U.S. Navy clearly saw the technical maturity of their offering. Having the experiment highlighted in *Aviation Week & Space Technology* was a boon to the campaign for socializing a key set of messages with the Navy and perhaps throwing the competitors a bit off balance.

In addition to employing the test bed, another element of Head Start was investing in the building of a BAMS Global Hawk development test article, or "white tail" aircraft, with some of the Navy-unique capability built in. This would be done on company funds upon winning the award and would be in advance of the system development and demonstration phase. These two key elements comprising the Head Start initiative were designed specifically to reduce risk and create real technical discriminators for the Northrop Grumman team by addressing the value proposition to the customer.

Two more elements of the strategy were essential to supporting the win. The first involved committing to a much earlier initial operational capability—FY2012 instead of FY2013—in order to meet the BAMS objective requirement. With the threshold requirement set for FY2014, a strategy element was created to accelerate the initial operational capability by two years—another bold decision given the risk involved, but one that garnered favorable Navy reaction. Mitchell and Wood were confident of their team's ability to manage this accelerated schedule based on the current Global Hawk production schedule. This earlier capability would save the Navy significant dollars and avoid substantial cost.

The other of these two elements of the strategy was to perform final assembly of the aircraft at a technically qualified facility that would not

---

Aviation Week and Space Technology got a behind-the-scenes look as Northrop Grumman demonstrated its prototype BAMS sensor on a test flight Aug. 17, during which operators collected images and data on specific targets. Information was transferred to a prototype mission control station (MCS) located in Hollywood, Md. The demonstration forms a key part of Northrop Grumman's Head Start risk-reduction effort for the $2-billion BAMS program.

—GUY NORRIS and AMY BUTLER, "Head Start: Gulfstream
11 Risk-Reduction Flight Tests Key to Northrop Grumman's Strategy for
BAMS," *Aviation Week & Space Technology*, September 24, 2007

---

only enable a lowest cost option but also would be attractive for the Navy for securing political support due to the jobs that would be created in those legislative districts. The final assembly location for Global Hawk at the time was in a government-owned facility in Palmdale, California, which not only cost more, but also was dominated by an Air Force presence for the Global Hawk assembly and other programs. Many, particularly within the U.S. Navy, viewed Global Hawk as a U.S. Air Force airplane. Northrop Grumman was continuing to spend considerable time and investment in marinizing the aircraft to make it a uniquely Navy UAV. The team became very focused on a unique standalone final assembly location for the BAMS offering to show a wholly Navy-centric customer approach for this effort.

Although some of the elements of this win strategy matured over time, the fundamental points remained constant and pertinent to the capture throughout the bid process. Its eventual success proved it was a compelling win strategy, but that was made more evident during the competitive process by virtue of its initial thoughtful objectives and tactical elements aimed at teasing out discrete, measurable discriminators, drafted by a tightly bonded and organized group of experienced proposal experts in competitive government procurements. Its success highlights by comparison that poorly thought-out win strategies and customer value propositions can negatively affect the probability of winning the competition or create vulnerabilities that may be countered by the competition. It cannot be overstated how important to the success of the strategy was the execution—meaning close adherence to the strategy and its win themes throughout. Only an experienced leadership team that maintained all participants on a steady course and did not allow surprises or competitor actions to lure them off in an unproductive direction could enforce this.

Wood's overriding focus in the bid response was to ensure that win themes, as derived from the win strategy, were fully incorporated into the draft proposal. He developed a win theme roadmap document that matched each volume of the proposal to its relevant themes as it aligned with section L of the Navy's early draft RFP, containing instructions from the government for formatting, organizing, and submitting the proposal. As the proposal manager, Al Hutchins's experience in defining and maturing win themes was immensely helpful in this process. Each proposal author received these win themes with their writing assignment, whether technical, management, classified, or international volumes. Wood's goal was to weave the win themes into the proposal narrative early in the writing process where they could ultimately be matured as an integral feature of the proposal when it went into the later stages of review.

Proposals in the government contracting and defense business are very complex but, in their simplest forms, are stories about an offering and what its value proposition is for the customer. Thinking of a proposal as a story is critical to the continuity and readability of the overall narrative and to making it compelling for the customer. Engineers and technical people on the proposal team often forget this, and their specialized academic training frequently impedes clarity in their writing. As the campaign leader, Wood viewed one of his critical functions as being the spokesperson and liaison for the customer, ensuring that the company's engineers infused their drafts with themes, even language, that resonated with the customer and conveyed that the team understood and shared their vision for BAMS. He directed the construction of a large sign for the site's main conference room in the proposal center that read, "Remember, we must win it first." This was designed to remind the engineers in particular that adding technical detail beyond what the customer asked for would only detract from the impact of lean, precise descriptions that directly responded to the RFP inquiries. Greater levels of detail could come later after first properly and succinctly answering the RFP and winning the competition. Al Hutchins was also a critical advocate of this approach.

An area of particular focus for Wood, and one that formed the most visible outreach for the capture team outside of the sector and company, was the marketing and communications effort for the campaign. Wood pulled Tom Twomey into this area after the short tenure of Wood's Melbourne-based hire. Some of the simplest components of marketing communications were the customary token mementos such as golf shirts, pens, and other paraphernalia reserved for handouts to industry people and, if possible, Navy customers. The effect is immediate recognition in whatever venue they appear: to provide a reminder and visible connection to the Northrop Grumman BAMS concept (although most direct Navy customers would not take specific contractor items while the competition was in progress). Little tangible benefit is derived from this activity other than brand recognition and visibility around Navy circles. However, Wood and Twomey put a lot of thought into how they could design these small, low-cost giveaway items to be more useful in conveying a message shaped by their win themes. Several marketing and communications initiatives were immediately employed that were quite unique and provided real information on the benefits of the Northrop Grumman BAMS offering. The two were confident that the PUMAS study showed their offering, by comparison to the competitors, in a favorable light.

The team converted the PUMAS study technical summary document in the technical maturation phase of that effort into a marketing brochure that could be handed out to anyone—customers, industry, or legislative staffers on Capitol Hill. They agonized over whether the customer would authorize this, since it was really their original product document, but a decision was made to go ahead with it. The engineers on the team fought it somewhat, but the business development people on the BAMS capture team prevailed, citing the adage that it was "better to beg forgiveness after the fact than ask permission before." This color brochure was a comprehensive document explaining in graphic detail why the Global Hawk BAMS offering was the most compelling solution for the requirement. It was replete with Northrop Grumman win themes and the ghosting of the competitors' alleged attributes. Specifically, it depicted competitive strengths of the Global Hawk BAMS offering—space, weight, and power, range, ETOS, sensor capability, and so on, in appealing color graphics. By properly wording the win themes and suggesting how the competitors' solutions lacked these strengths, the team was ghosting the competitors and subtly pointing out their weaknesses as well. This marketing brochure went to a long list of exclusively selected recipients.

An important adjunct to the messaging campaign was the full employment of sector field office representatives. This included the Patuxent River office for engagement with NAVAIR functional departments—engineering, logistics, contracting—and the program offices; the Norfolk, Virginia–sited office for engagement with the maritime patrol community; the North Island, California, office to reach the staff of commander, Naval Air Forces—chiefly responsible for aircraft systems readiness; and the

Washington, DC, office, where company representatives, now including Ernie Snowden, coordinated the sector's representation in CNO staff offices, with Navy labs, and at Navy events where a concentration of flag officer influencers and deciders were always present. Coordinating with the BAMS capture team for the desired messaging, a full-on effort was maintained to equip all of the integrated systems sector business development field representatives (and frequently some from the other sectors) with tabbed handbooks containing key win theme messages targeted to specific flag officers by name and position. Scheduled events such as the Surface Navy Association Convention, Tailhook Reunion, and Navy League's Sea-Air-Space Expo were the typical venues where the flag officer engagements were arranged to appear coincidental, but wherein the company business development reps were charged with delivering the key messages and obtaining as much feedback as could be shared. These messages centered on the positive details of the Head Start risk reduction effort, most especially the maturity of MFAS radar on the company's flying test bed aircraft and the state-of-the-art domed BAMS podule in Hollywood, Maryland; milestone accomplishments in the GHMD program that had particular relevance to the BAMS effort; and the anticipated performance of the preferred system concept, eventually to be designated the RQ-4N, against Lockheed's and Boeing's performance and as measured against threshold and objective performance requirements.

Another initiative the team embarked on was the creation of a trifold handout that summarized Global Hawk's key attributes and offered a cogent, logical rationale for why the Navy should select Northrop Grumman's offering for the BAMS solution. Postal and email addresses for every customer the team had or expected to have direct or indirect contact with were logged, and with Mitchell's approval, each was sent the trifold in a blanket mailing. The team was concerned that this might go against the program office's restriction on communication, unduly influencing the decisionmakers in the competition, but since the mailing would go to everyone possibly known in the Navy acquisition world—CNO staff, the fleet, and the Hill—it would be considered just a blanket marketing effort with no specific targeting of any one person. There was some concern when the program office pushed back after the mailing, but that quickly died down after they were told that it was harmless and went to everyone as a Navy-wide distribution.

The third major effort under marketing and communications was placement of a uniquely designed advertisement in the Naval Air Station Patuxent River base newspaper. It ran as a full-page advertisement inside the back page on a recurring basis until its initial impact was likely lost once the proposal was submitted. The advertisement generated as much internal team discussion about its efficacy in influencing the customer as it might have stimulated with the intended customer target group itself. Many questioned the real worth of the ad in generating customer reaction, even though the practice is commonly employed as a marketing and communications tool in large competitive campaigns. The occasional use of a full-page advertisement in the *Washington Post* (which can cost more than $100,000) or an advertisement on a local radio station was often debated regarding cost and the return on investment. Neither venue was pursued, but Northrop Grumman's KC-X Air Force tanker capture team had a different perspective and did pay for these products while competing against Boeing, which was acknowledged as a behemoth with deep pockets for flashy newspaper or radio space. Gauging the benefits of these marketing efforts in a campaign is difficult, but key customers were known to be reading or hearing these ads and could be unconsciously influenced toward a Global Hawk–based solution for BAMS.

Prior to the development of the marketing materials, considerable thought was given to what to call the aircraft. The team searched for a unique name for the Global Hawk BAMS offering that tapped into a legacy from an earlier time, to an aircraft with a recognizable history and reputation in maritime patrol or ISR, preferably one flown by the Navy and produced by one of the corporate antecedents of Northrop Grumman. Once the Global Hawk BAMS offering was ultimately selected, the Navy would officially confer its own name on the unmanned aircraft, but in the meantime, a popular name was needed that people could associate with the system—something that was representative of Global Hawk and the marinization of the system for the Navy.

Global Guardian, which borrowed "Global" from Global Hawk and "Guardian" from a post–World War II Grumman airplane designed for

Navy maritime surveillance, was the initial choice. Wood undertook significant research with a corporate attorney on trademarking this name, but the process was too hard. Some on the team and in the company did not like the name and resisted it, which became a challenge and detracted from the original intent of the naming exercise. Ultimately, Wood and Mitchell decided to drop Global Guardian and came up with something very simple, RQ-4N. RQ already had precedence as a standard military designation for a reconnaissance unmanned aerial system, and the 4 signified it was the fourth in the series of that type produced by Northrop Grumman. The N was added for Navy. It was simple and noncontroversial, and it stuck for the tenure of the capture and the proposal.

One of the principal aspects of the strategy involved the contract on GHMD. It had already been formalized under contract with the Navy for two U.S. Air Force Block 10 aircraft in demonstration phase for the potential BAMS program. The Air Force Block 10s were outfitted with a Raytheon side-looking mechanically scanned array radar and a legacy electronic support measures system, which was not optimized for the maritime surface search ISR mission. But it still provided a great platform for the developing concepts of operation for the BAMS mission centered on the Global Hawk aircraft.

This effort to get a funded contract for the demonstration phase using Global Hawk was a huge coup for the company. What GHMD could now do in the midst of the BAMS competition was provide the pretext for being continually with the Navy customer showing them how Global Hawk was best positioned for this mission of maritime surveillance. This was an enviable position for any contractor and was a big part of the strategy to shape the customer toward a Global Hawk solution for BAMS. The advantage was so significant that the Northrop Grumman team sometimes wondered how Lockheed Martin, General Atomics, or Boeing would strategically counter GHMD's presence in the BAMS competition and proposal. GHMD was truly the "elephant in the room," but fortunately it worked in Northrop Grumman's favor and to the disadvantage of the competitors.

As it turned out, entering 2008 with the draft RFP issued, those competitors did not have much to worry about; both airplanes were mechanically down for parts supply and were literally gathering bird droppings on their wings inside the hangar or parked on the flight line. What seemed to be a competitive advantage with GHMD had the potential to become a distinct liability, especially if the competitors fully appreciated its problems and began ghosting the more negative aspects of GHMD readiness. As this liability was analyzed, the team began to realize this could have the effect of ruining the competitive advantage and in fact causing them to lose the competition in the end game. Since the two GHMD Block 10s were from the U.S. Air Force inventory, and given the parochialism of the two military services, there was no urgency or any sense of priority evident in the Air Force for expediting spare parts. The Navy program office struggled with the issue of obtaining parts for the aircraft, and Northrop Grumman, stretched thin in other program issues, was not doing much to help. It seemed to be at a logjam, and little progress was being made to resolve the problem.

During the blue team strategy reviews and in front of the consultant review boards, the issue was highlighted and advice was given on what to do and what the senior advisors thought the impact would be. Wood briefed this issue at one review board to an assemblage of retired two-, three-, and four-star flag officer consultants and industry experts. They were in near-unanimous agreement: if Northrop Grumman did not fix this problem as soon as possible, its efforts were doomed to fail on BAMS, and the probability of winning might plummet. This wakeup call immediately energized the capture team to make this the top issue. This also got significant visibility up to the sector and corporate leadership. Sector president Scott Seymour decided to assign vice president Carl Johnson, an experienced Global Hawk program manager, to go to Naval Air Station Patuxent River and work the issue full-time until the aircraft were airborne. Johnson had run the Global Hawk program from the Rancho Bernardo site for several years and knew the airplane and its systems. Unfortunately, expectations on what he could deliver and how quickly became enmeshed in some internal company politics and became a concern for the capture team, given the heightened visibility this issue now had. But as a well-seasoned program manager and an excellent engineer who knew the technical aspects of the system well, Johnson was able to highlight and bring into focus the more problematic issues. After some discussions between Scott Seymour and Bob Mitchell, Johnson complied with his direction from Seymour and headed to southern Maryland to work the problem.

Within four months, Johnson was able to get one of the aircraft in the air and operating on some demonstration flights. Eventually, he got them both flying. This was a significant turn in the capture effort for BAMS: a potentially ruinous trend for GHMD, and for Global Hawk BAMS by association, had been righted and was now moving in a positive direction. If the problem had not been corrected, it would have seriously undermined a critical strategy element that relied on obtaining green color ratings in past performance ratings in the BAMS proposal evaluation. Since GHMD was essentially the same air vehicle as the BAMS offering, its performance was critical to the credibility of the Northrop Grumman BAMs offering. The Navy's observation of GHMD in its own back yard could make or break the Global Hawk BAMS concept in that evaluation category and possibly skew the evaluation in other categories as well.

With a disciplined daily application of program management fundamentals, Johnson seemed to be getting this difficult problem back on track. His stock rose in the company and would soon yield improved career prospects for him, eventually following Mitchell in the role of BAMS capture manager after the proposal was submitted to the Navy. So a major crisis was averted, and Northrop Grumman leadership finally focused on this problem and worked to resolve it. They had procrastinated for some time on resolving it, but the blue team review and retired flag officer recommendations were a critical juncture in the overall success of the capture. Of several defining moments for the capture, this could certainly be labeled the most crucial one, for without a resolution, the ultimate win and success for Northrop Grumman might not have occurred.

ADHERING TO THE WIN THEMES

Once the proposal was dropped on the Navy, an exchange began with Navy evaluators to clarify statements made in the proposal and to amend particulars in the way certain requirements were addressed. The common currency of this interaction was the government written evaluation notice (EN). The notices were numerous—numbering in the hundreds—and normally described a discrete issue the Navy evaluators had with proposal wording or the concept proffered that typically ranged from incidental and trivial to more serious and possibly disqualifying. The EN process proved important to the overall proposal process both by highlighting potential red flags that could be addressed in an EN response and by opening a path into the evaluation process to expand upon or reinforce key win themes. During the EN process with the Navy evaluators, an opening to write more thoroughly about the Head Start effort materialized, permitting greater emphasis on risk reduction efforts. Al Hutchins took the lead in helping to write and edit this critical proposal change. It was particularly gratifying to Mitchell and Wood that what many saw as a risky, bold initiative now seemed to get the serious consideration in the EN process that was originally hoped for.

One example of how the EN process was used effectively to reinforce win themes showed once more that well-thought-through ideas, observed throughout the process and administered by a steady hand from the campaign leadership on down, can avert potentially disastrous setbacks. The Navy evaluators sent an EN that drew attention to the proposed electro-optic/infrared sensing system that went by the trade name Night Hunter II as an external program dependency. The system was under development concurrently by the Navy P-8 MMA program, and the stated plan in the BAMS proposal was to adopt the system once matured by the P-8 program, thereby avoiding some of the development cost that might be assessed against the Global Hawk BAMS configuration. The proposal highlighted a premise that Night Hunter II's performance, historical programmatic support and funding, provided an extremely remote chance of the program being cancelled or delayed in FY2008.

The day after receiving that EN response, the Navy P-8 MMA office cancelled the Night Hunter II contract. This threw the Northrop Grumman BAMS team into a quandary, concerned that this was of sufficiently high-risk importance to their proposal that it could signal a nonacceptable score. They referred back to the basic win theme, which was to declare Northrop Grumman to be a proven prime system integrator in response to the original RFP's demand for "responsive management." By leveraging

DOD investment in a proven sensor system, the proposal was able to offer a low-risk program. Of course, examples were cited to substantiate the claim of responsiveness, based largely on consistent application of, and expertise in, systems engineering management and efficient supplier base management. Now, the assertions were being tested in real time.

Within days, the team was able to re-engineer an accommodation for an alternate electro-optic/infrared system, the multispectral targeting system–B (MTS-B), and to broker a vendor arrangement with Raytheon for supplying the system. In response to their win theme claim of being a proven system integrator, they could now state through the EN response that they invoked their system engineering process to assess recent stop work and assessed that a change to the baseline offering from Night Hunter II to MTS-B was the best path forward. The team then worked with Raytheon to fully reflect their selection as electro-optical/infrared provider. To assure the evaluators that DOD investment could be leveraged to obtain a proven sensor system, the capture team discussed that its evaluation of the MTS-B established that the sensor met the sensor requirements and had a sizable margin within Northrop Grumman allocations of cost and schedule.

In the first round of discussions after proposal submission on May 3, feedback of the Navy's initial evaluation indicated that one component of assessed technical evaluation had much to do with the statement of objectives included in the RFP stating that each BAMS orbit provide (with no more than three aircraft aloft at one time) continuous surveillance for twenty-four hours in each of seven consecutive days with an effective time on station of no less than 80 percent. Similarly, the performance-based system specification in the RFP required that the proposed aircraft system maintain an 80 percent threshold and 95 percent objective ETOS within 168 continuous hours at 2,000 nautical miles from the operating base. In assessing the Global Hawk–based BAMS offering, the Navy determined that an ETOS of 96.2 percent was achieved for a three-aircraft orbit, well exceeding the objective requirement. Not revealed then but disclosed later was information that the Navy evaluators determined that Lockheed Martin/General Atomics' Mariner aircraft, powered by its single turboprop operating at 2,000 nautical miles from its base, yielded an ETOS for a four-aircraft orbit of 84.6 percent, and Boeing, with an undisclosed number of orbiting aircraft, at 92.8 percent.

Partially on the basis of these assessments, the Navy determined all three competitors to be in the competitive range and, after dialogue with each, asked for final proposal revisions. In the final analysis, determination of an overall technical rating would add to ETOS other factors having to do with design approach that included scoring for open systems architecture, mission performance as a measure of sensor capability, due regard (separate forward-looking collision avoidance radar), and growth potential in the areas of space, weight, and power, all areas that were addressed specifically in the Northrop Grumman win strategy.

During the ENs, a deficiency on cost risk surfaced, with a red rating despite efforts to get this right with the customer. It showed that the cost bid was almost 100 percent out of bed with NAVAIR cost evaluators. At a senior-level conference call hosted in the Washington, DC, office, the corporate leadership joined Mitchell and Wood in conferring on what to do before final proposal submittal (specifically the final price), as the Navy was asking Northrop Grumman to *raise* its price. Such a response could lead to a very unsettling place, given that there still was a strong possibility that lowest cost could determine a winner. The team members believed they had truly shaped it to be a best-value procurement, but this was still unpredictable. There was no unanimous consensus among Northrop Grumman leadership on what to do, and much debate took place, which demonstrates the business complexities of a multi-billion-dollar proposal offering such as BAMS. In the end, the price was raised, as requested by the customer, but not as much as they desired. It proved to be a winning decision.

Competitive captures of this magnitude always live in a state of chaos and, if done correctly, foster various levels of paranoia on how the capture team is doing. The chaos is a product of the massive amount of work and cost required to ensure the proposal team is driving toward a winning solution and ultimately a top-notch proposal product, all while the clock is ticking and a hard-and-fast submission deadline is nearing. The underlying paranoia is generated around the fear that something critical might have been overlooked, that the competition is doing something unknown to shape the customer, or that the customer harbors some ill-conceived notion of the offering or the team that could hurt the chances of winning. Things hinted at or said in the public domain or found out through an industry network could signal a missed critical item. In these cases, the team needs to react and adjust to the information, although some of it could prove to be all or partially correct or completely wrong.

Inside Northrop Grumman's BAMS capture team, chaos and paranoia involved several critical areas: concern over the late start in getting organized for receipt of the RFP and the team's lag time with the proposal development and the steps involved, such as locking down the technical baseline; internal disputes over the correct technical offering, especially decisions on key capabilities such as the radar; changing government requirements late in the game; worry about biases held by the Navy program manager and program team that could be a disadvantage; and internal discord between the Bethpage and the Rancho Bernardo parts of the team. Each of these detractors was unique in its own way but, taken together, became a daunting challenge to resolve in the limited time available until submission of the final proposal product.

At least two days prior to the BAMS award announcement by the Navy, the Northrop Grumman team was notified by corporate legislative lobbyists on Capitol Hill that the Navy was briefing some of the professional staffers for the defense committees. Wood's full attention was on this, trying to get some inkling of the award decision by working closely with Diane Harper, a lobbyist at the Rosslyn, Virginia, corporate office. She uncovered some intelligence that Northrop Grumman was the likely winner and should anticipate the call from the Navy confirming selection. Wood had previously directed his business development group to build and preset an advanced website to launch marketing materials to the widest possible dissemination the moment that a win could be confirmed, so he was getting ready to push that button. He waited in the Bethpage proposal center with several of the BAMS capture team, including Carl Johnson, who was taking the role of BAMS capture executive from Bob Mitchell.

Rear Adm. Tim Heely, the unmanned aviation and strike weapons program executive officer, initially called integrated systems sector president Scott Seymour with the news that would ignite tremendous excitement in the proposal center and in the company's unmanned

---

The Department of the Navy announced today that the Northrop Grumman Corp. has been awarded the system development and Demonstration (SDD) contract for the Broad Area Maritime Surveillance Unmanned Aircraft System (BAMS UAS). The BAMS UAS will be developed using Northrop Grumman's RQ-4N platform. "This announcement represents the Navy's largest investment in unmanned aircraft systems to date. This is a significant milestone for the BAMS UAS program, concluding a deliberate and meticulous source selection process that adhered to stringent Federal Acquisition Regulation and Naval Air System Command source selection processes and documentation requirements," said Capt. Bob Dishman, program manager for the BAMS UAS program.

—Department of Defense, "Navy Awards Northrop Grumman
Unmanned Aircraft System Contract," April 22, 2008

---

systems unit in Rancho Bernardo. Within minutes of that call, the Navy's principal contracting officer and the BAMS program manager rang on the landline. They broke the news that Northrop Grumman had won and that the Navy customer was very pleased and excited about it. All of those months of paranoia and uncertainty suddenly were forgotten in a jubilant outburst of celebration.

After sharing congratulations with the NAVAIR contracting officer, she faxed a completed form for the $1.16 billion cost-plus-award-fee contract for the system development and demonstration award. The BAMS contract would include funding for two unmanned aircraft, a systems integration lab, and two mission control systems: one primary base system and a second system designed for a forward operating base. Wood pulled the contract off the fax machine to scan it quickly to ensure it said what they were told on the phone; sure enough, there it was in writing. All were somewhat stunned by the euphoria of the moment and the successful culmination of years of hard work that it immediately conveyed. Wood immediately contacted Tom Vice on the award, and soon thereafter Wood and a handful of others started planning the win party in the Bethpage environs. The buoyant celebration of the moment lasted only for a few days—until it became evident that one of the competitors would be protesting the award.

CHAPTER 10
PROTEST

Two Northrop Grumman–declared must-wins for its integrated systems sector remained undecided heading into 2008: KC-X, the Air Force's next-generation aerial refueling tanker, and BAMS, the Navy's unmanned aircraft system for maritime ISR. For the KC-X opportunity, Northrop Grumman led a team that included EADS, European manufacturer of the candidate large-body KC-30 aerial tanker. Northrop Grumman's win strategy for the Air Force's KC-X tanker competition bore some resemblance to the BAMS win strategy but also differed in key ways. A missionized tanker derivative of the commercial A330, the KC-30, was already in service with the Royal Australian Air Force. The KC-30 had been modified for the mission from a fairly modern inventory of Airbus large-body aircraft in commercial service around the globe and featured defense systems, precision fly-by-wire, and the ability to integrate a sophisticated militarized communications suite. In a technology risk-reduction demonstration resembling the BAMS Head Start initiative, Northrop Grumman and EADS showcased a functioning fly-by-wire remote aerial refueling operator console in actual airborne refueling of fighter aircraft, and the team was constructing the equivalent of a tanker white tail that would demonstrate further commitment to reducing high-risk items. And just as the BAMS internal win strategy mandated a site selection that was most beneficial to its customer, the KC-X win strategy similarly mandated domestic U.S. production in Mobile, Alabama, to allay concerns over offshore manufacturing. In the leadup to KC-X award, Northrop Grumman's tanker campaign team had determined that shaping the evaluation criteria to be based on best value was central to winning the competition against Boeing's smaller, lower-cost KC-767 derivative tanker. The strategy proved effective in delivering a must-win for Northrop Grumman at the end of February 2008, but the KC-X award was ultimately overturned by Boeing's surprisingly aggressive political protest.

Once Boeing formally filed its protest of the KC-X award in mid-March 2008, the U.S. Government Accountability Office (GAO) took center stage to adjudicate the efficacy of the Air Force award. From the start, GAO made clear that it was not reviewing the relative merits of the two competing aircraft but was investigating only whether the Air Force complied with statutory and regulatory mandates governing the federal procurement process. Behind the scenes, however, evidence was undeniable that Boeing was pulling out all stops and calling in political markers to ensure that GAO came to the right conclusion and, short of that, to put in place the legislative measures that corrected this potential loss of their franchise. The intensity of the Boeing protest campaign caught Northrop Grumman still basking in the glow of its KC-X win and slow off the mark to begin its counter strategy.

Since the KC-X tanker competition was initially awarded before BAMS, the BAMS capture team was in a unique position to learn from the many challenges that the Northrop Grumman tanker team experienced. With that tanker award to Northrop Grumman, which surprised many inside both DOD and industry, Boeing started an almost immediate "scorched earth" campaign to do whatever it took to execute a protest and launch a marketing campaign to reverse the award. The Boeing campaign, targeted at senior leaders in the Pentagon and Congress and carried out through its public relations team, was on the level of something the industry had never seen before. The BAMS tanker capture team watched incredulously from the sidelines, recalling the countless hours they worked to defend themselves and support their U.S. Air Force customer's selection. They were completely unprepared for the intensity of the Boeing attack, particularly Boeing's enlistment of their congressional representatives in the process.

Since Northrop Grumman was also competing against Boeing as well as Lockheed Martin/General Atomics in the BAMS competition, the BAMS capture team began to put together a comprehensive defensive plan to prepare for a protest on the award. With the understanding that the tanker competition represented a live-or-die franchise program for Boeing, the loss of KC-X was assumed to be a much bigger deal for Boeing than BAMS could be. But the BAMS team nevertheless wanted to be prepared for a potential protest coming from any direction and began to strategize for an attack from either Boeing or Lockheed. Bob Wood pulled together the process and action plan for how to handle a protest and deciding who would do what upon learning of a protest from the Navy customer. He examined the tanker team's marketing strategies and lessons learned and also how they engaged their Air Force customer during the one hundred days of protest and evaluation by GAO, the agency charged with providing an impartial forum for the resolution of disputes surrounding government contract awards.

The BAMS campaign team, under Wood's guidance, developed a concise series of actions and responsibilities for their leadership team, actually creating a new phase in the sector's published business acquisition process titled "Post Award and Protest" that all capture teams were subsequently required to address in their planning. This was never documented before in the business acquisition process simply because no one had ever been through this level of intense post-award competition before. The intention was to stand up this effort in preparation for a protest upon award of the BAMS contract to Northrop Grumman, calculating that the competitive environment suggested they should expect at least one competitor to protest. After working through a competitive analysis of which company would likely protest, they determined it to be the Lockheed Martin/General Atomics team.

The award was clearly a must-win for both of these companies. A Lockheed Martin loss could be the final nail in the coffin for its Eagan, Minnesota, site that was the legacy corporate domain of its Navy maritime surveillance business. If General Atomics lost, the hoped-for expansion of the Predator UAV aircraft into the Navy would be abruptly halted, ending its efforts to expand its customer base in DOD beyond the U.S. Air Force and U.S. Army. The maritime surveillance domain for its unmanned aircraft would be relegated to international customers only. Without a U.S. Navy customer, those countries desiring this maritime surveillance capability would be hard pressed to have confidence in a system that the U.S. Navy had decided not to procure. Since the majority of the Earth is covered by water, many countries with a coastline desired a maritime surveillance capability, and an unmanned system would come with lower procurement and support costs for them.

Losing the BAMS competition represented a significant business loss for General Atomics. Most concern centered on Lockheed Martin/General Atomics, but they had observed Boeing's conduct on KC-X and were not going to take for granted that Boeing would not join in the protest. They prepared at length for a protest and briefed Northrop Grumman leadership to ensure they were satisfied that any missteps by the tanker team would not be repeated by the BAMS team.

The Navy announced its BAMS award to Northrop Grumman on April 22, barely five weeks after Boeing's protest was filed on the tanker award. The BAMS team had the benefit now of hearing the Navy's summary evaluation and award rationale in its debrief, which was offered to all the competitors but tailored for each so as to not reveal key proprietary information from the other competitors' bids. In the Navy debrief, they were able to determine that once the final proposal revisions had been submitted, their evaluation began to clearly show some separation between the competitors. Even though all competitors were judged to be highly satisfactory with minimum risk in the technical ratings, Northrop Grumman's and Boeing's proposals were evaluated as having a strong advantage over Lockheed Martin's overall technical approach. The Navy evaluators' most important factors—ETOS, open systems architecture, mission performance, and room for growth in SWaP—carried the most weight in this regard. NAVAIR evaluators noted that Northrop Grumman's Global Hawk–based BAMS concept showed significant strength in the emphasis area of SWaP for an ability to add future capability increments without breaching ETOS threshold requirements. By contrast, it was learned later that the Lockheed Martin team's Mariner concept was evaluated as showing significant weakness in its design approach for failing to provide a validated computer model of the performance of its proposed engine.

The Navy evaluations began to show substantial differences between competitors in their assessed past performance and experience. Relative standings of the competitors as determined by the Navy source selection advisory council showed the following:[2]

NAVAIR Evaluation of Final Proposal Revisions

	Northrop Grumman	Lockheed Martin	Boeing
Past Performance	Moderate Risk	High Risk	Low Risk
Experience	Very Low Risk	Very Low Risk	Low Risk

Source: Gary L. Kepplinger, "Decision in the Matter of Lockheed Martin MS2 Tactical Systems," U.S. Government Accountability Office, August 8, 2008, 4.

In assessing Lockheed Martin's past performance, the advisory council zeroed in on the poor record of its subcontractor and teammate, General Atomics, for the unmanned aircraft. The Navy noted that customer feedback was remarkably consistent across multiple contracts in two other military services for Predator-related work for poor performance in managing workload and in executing systems engineering and systems integration tasks. Singling out that team's Mariner prime Lockheed Martin, the council noted that the company had difficulty adequately staffing its Po Sheng command and control system work for Taiwan, prompting concern that Lockheed Martin would encounter schedule delays requiring technical trade-offs.

Following the Navy award to Northrop Grumman for the BAMS contract, the team immediately launched their post-award and counter-protest effort. They knew this was not over and worked to assess what competitive intelligence could be gained as to how the award was being dealt with by both competitors. There was little in the public domain that would indicate protest activity. Except for the standard press releases about being disappointed and looking forward to the debrief from the Navy customer before taking their next steps, Lockheed Martin and Boeing were being quiet about it. At the same time, Wood and Twomey worked with the Northrop Grumman communications and customer relations team to plan the BAMS win party. After significant planning, a large corporate-level win party was held on May 6 at the Oheka Castle, a historic hotel in Huntington, New York. The capture team and the company had so much to celebrate after many years of hard work and persistence on designing, maturing, and ultimately providing the BAMS winning solution for the U.S. Navy.

The Navy took a while to schedule the debriefs, finally informing Northrop Grumman that it would be last. The debriefs were conducted in a professional but open manner in the program manager's spaces at Naval Air Station Patuxent River. The meeting was positive and marked by frank and optimistic discussions with the program manager, Capt. Bob Dishman, and his team about their selection of Northrop Grumman and the RQ-4N. Northrop Grumman was informed that it did not submit the lowest price in this best-value competition but that in the evaluation of all the proposals, its offering gave the Navy the best-value capability for BAMS. This echoed what was revealed in the debrief: a robust technical offering—that included better SWaP-C, superior endurance, range surveillance coverage, and sensor superiority—combined with strong marks in past performance and experience more than offset a higher price relative to the competitors.

Following the debrief in the Navy's conference spaces, a discussion ensued with Captain Dishman over the decision to select the MFAS radar. He was almost incredulous that the winning team would have struggled with that choice, given how pleased they were with the radar solution. The team told him that they struggled principally with the cost and its ultimate impact on price. He responded that it was the state-of-the-art technical solution and it was the right one to offer as the centerpiece of the sensor system. The Navy team revealed in the debrief that they had rehearsed all debriefs in detail to ensure that nothing incorrect was said and that they were prepared to field any questions from the contractors. They were being very careful to try and mitigate a potential protest from one of the competitors.

Leaving Patuxent River that day, the team was ecstatic over their win and what seemed to be the end of an intense competitive journey. But since the clock was still running that gave any competitor a window for protest, a nagging concern remained about one materializing. Within the next five days Wood organized the schedule of meetings with all key suppliers on the BAMS program at a Washington, DC, hotel. The intent of these meetings was to fill everyone in on both key aspects of the debrief from the Navy and on important next steps that were planned for program kickoff. The engineering and manufacturing development phase would soon begin, signifying the next step in the formal DOD acquisition framework and the official initiation of a major defense acquisition category program. The purpose of this next phase was the development of the BAMS capability—really, the fulfillment of the proposal offering. Coming after a major milestone review (milestone B), it consisted of two distinct efforts: integrated system design and system capability and manufacturing process demonstration. But it also put in motion the preparation for a critical design review assessment at the conclusion of the system design effort. The overall goal of this phase was to complete the engineering development of the BAMS capability and proceed into production and development.

The leadership of the supplier firms assembled in the conference room were in high spirits as the Northrop Grumman leadership team began a series of briefings. As Bob Wood was briefing, he noticed a flurry of activity in the back of the room. The leadership team, now including Carl Johnson as the capture executive, was growing increasingly animated, giving off an air of worry. Wood could sense that something was wrong. After his prepared remarks, he called for a break before continuing with the next briefing. As he joined the discussions at the back of the room, he was told that the Navy team had phoned to inform Northrop Grumman that the Lockheed Martin team was submitting a formal protest of the award. The Navy representatives confided that they did everything they could to convince the Lockheed Martin leadership not to file a formal protest, but they were ultimately unsuccessful. Since the protest was submitted within ten days of their formal debrief, the Navy would have to bring the program to a complete halt until the protest was resolved. The Government Accountability Office then had one hundred days to formally review the specific aspects of the protest. This news cast a pall over the conduct of the meeting. The supplier company representatives were informed, but scheduled presentations were finished before people departed for their home offices.

Boeing had decided not to protest the award after learning in their evaluation debrief that they were much more expensive than Northrop Grumman in their total price. A year after the protest was resolved, Wood would learn from a retired Navy flag officer involved in the evaluation that Boeing was over a billion dollars more expensive due to their complex radar and avionics solution. The Navy was enamored with the unique offering but could not afford it. In this case, Boeing apparently miscalculated when trying to find a winning balance between a highly technical, innovative, even elegant solution, and the Navy's available budget to pay for it.

COUNTERING THE PROTEST

The next day, Northrop Grumman's BAMS team launched into their protest strategy action plan. Having predetermined that Lockheed Martin/ General Atomics would be the likely protesting team, they had a marketing and legal course of action plan set up, in addition to a communication plan with the Navy customer. The team launched into high gear after fully notifying their leadership up to integrated systems sector president Scott Seymour and CEO Ron Sugar. They were prepared for Lockheed Martin to launch an onslaught of negative publicity and marketing communications, but into the first week, they were noticeably quiet about the protest except for a few general public comments. There was none of the scorched earth marketing campaign that was so characteristic of Boeing on the tanker competition.

The team realized that the different dynamics of the two competitions were greatly determining the forcefulness and intensity of the companies that were protesting. In the KC-X tanker competition, Boeing stood to potentially lose a forty-year incumbency in the mission of aerial refueling support to the U.S. Air Force. That same monopoly was not present in the BAMS competition. For one thing, this was a new procurement of an unmanned system in a historically manned mission for which there was no preexisting corporate footprint. In addition, Lockheed Martin was not Boeing, either culturally or in behavior in the public domain. Lockheed Martin CEO Bob Stephens had called Ron Sugar and told him that Lockheed would not behave the way Boeing did on the tanker protest, and that their protest would be handled in a professional manner. That was news to the team, but they implemented their entire protest strategy with no less vigor.

First and foremost was the introduction of general marketing materials on the website and elsewhere to emphasize why the Northrop Grumman BAMS team had won the competition and to present the merits of the offering to the U.S. Navy. In addition, the team retained a prominent Washington, DC, law firm known for its expertise in government contracting law and protests. The selected firm had a stellar reputation for helping companies to defend against a protest and for winning cases that upheld the original award to the client. What followed was a series of meetings with the BAMS team leadership in their law offices to educate the attorneys on the merits of the offering and to aid them in obtaining documents to support the Navy. It quickly became evident that the Navy was determined to shut down any avenue of communications with Northrop Grumman, as well as the protesting company, to preserve the integrity of the protest environment.

The Navy was also determined to ensure that once GAO ruled on the protest, Lockheed Martin would take no further legal action if their protest was denied, such as taking the protest to federal claims court. This venue is more formal and operates like a traditional court proceeding; federal judges hear the merits of a protest case, as contrasted with administrative attorney judges at the GAO level of protest adjudication. But taking a case this far is a last-ditch and painful process. The Navy did not want this to happen, as it invariably entailed further scheduling delays, lost funding, and lost capability for the fleet operators.

Even though the Navy had shut down communications, Northrop Grumman attorneys were communicating with the Navy attorney at NAVAIR. Soon after that dialogue opened, it became evident that the Navy attorney was inexperienced in protest law and that the Northrop Grumman attorneys would have to school her on how to handle this appropriately. This news was disheartening, for the last thing the company wanted to hear was the Navy was ill prepared for this protest. After a few weeks, the Northrop Grumman attorneys reported that things were improving with their case and with the Navy attorney's preparation to appear in front of the GAO judge.

In the first meeting, the company attorneys produced a copy of the redacted Lockheed Martin protest. Even though certain sections were redacted, enough information was revealed by reading between the lines to surmise the key grounds upon which the protest was built and to use it to prepare a defense for the Navy customer. The attorneys would then get these documents to NAVAIR without being in the middle of the communication. Lockheed Martin's grounds for protest mostly centered around their total evaluated price, which was lower than the Northrop Grumman price. Even though the Navy clearly announced that the competition would be evaluated as a best-value competition, Lockheed Martin was adamant that it should have been awarded the contract because of its significantly lower total price. Principal among many reasons for this was the fact that Global Hawk was a larger aircraft and more robust system. Navy evaluators factored in an assumption: airplanes are essentially purchased by the pound, so the more aircraft structure, materials, and weight, the higher the cost and ultimately—with fee added—the more expensive the price. The magnitude of any price difference in the respective bids, and the degree to which the Navy adhered to its best-value condition, was a concern throughout the competition. But what became clear was that the Navy, in its protest defense, supported the award at a higher total evaluated price largely based on the poor past performance of General Atomics.

The recent history of the General Atomics program management and execution of Predator and its derivative unmanned aircraft for the U.S. Air Force and U.S. Army was less than exemplary. The debrief from the Navy clarified that past performance and experience accounted for significant differences between evaluated proposals. But the significance of past performance, and the comparatively poor marks given General Atomics for managing its Predator programs, was more of a deciding factor. With Predator offered by Lockheed Martin/General Atomics as the centerpiece of its Navy BAMS proposal, the Navy evaluators were thus understandably reluctant to take on that poor past performance—especially since neither General Atomics, nor Lockheed Martin acting as prime on their behalf, could offer any U.S. Government documentation that supported improvement over those past performance problems. Throughout their protest, the core argument of the Lockheed Martin/

General Atomics case to overturn the award and rerun the competition was that their poorer evaluation ratings on past performance were unjustified and were not significant enough to merit the award to Northrop Grumman at a higher price. The next one hundred days were consumed with drafting, writing, collating, and transmitting multiple supporting documents to NAVAIR to refute the Lockheed Martin/General Atomics points and to support the Northrop Grumman attorneys and, in turn, the NAVAIR attorney.

From the moment of Lockheed Martin's protest filing and through late spring and early summer, GAO was running two parallel examinations of major award protests, one for the Air Force KC-X and one for the Navy BAMS, with information on progress and potential outcomes closely held. In GAO's view, this was purely a review of the integrity of the military services' evaluation and selection process and was not subject to external political influences. That said, it did not prevent the companies involved from exerting all the political influence they could muster to sway the outcome.

The recommendation from GAO to overturn the Northrop Grumman KC-X selection emerged first, in June 19. The office sustained Boeing's protest for seven specific reasons. However, two of those carried ominous portent for BAMS if the GAO rationale seeped into their BAMS assessment: if it was determined that the Navy departed from the solicitation criteria by giving undue consideration for exceeding key performance parameter objectives, and if the Navy improperly increased nonrecurring engineering costs and improperly used simulation models in determining those costs. As part of its KC-X report, GAO recommended that the Air Force "re-open discussions with the offerors, obtain revised proposals, re-evaluate the revised proposals, and make a new source selection decision, consistent with our decision."[3] Although not immediately obvious to the defense community at large, it was quite apparent to Northrop Grumman that its must-win KC-X opportunity had evaporated, done in by a combination of aggressive political maneuvering by Boeing and inexpert and inconsistent application of Air Force source selection criteria and protocols. The usual and expected pattern of GAO denying a protest and endorsing a service procurement selection had been, surprisingly, contradicted in the

KC-X outcome. The degree to which Boeing's relentless behind-the-scenes protest activity played into the GAO decision is hard to know after the fact.

With heightened interest, the industry waited the outcome of the GAO review of the BAMS award. The points of nonconcurrence that Lockheed Martin made central to its protest were unique to the BAMS selection process and did not, after all, bear much similarity to Boeing's charges on KC-X. But Lockheed Martin challenged the Navy evaluation as a misapplication of their ETOS because the Navy improperly based the Mariner analysis on a four-aircraft orbit. This allegedly ignored Lockheed Martin's EN response that they planned to base low-rate initial production cost and schedule on a five-aircraft orbit. The Government Accountability Office supported the Navy's finding that the subject EN was a cost EN, not a technical EN, and that the Navy had concluded properly that the proposal of thirty-three aircraft would not be sufficient to sustain full operating capability for the required twenty years. Moreover, when the Navy, in face-to-face discussions, informed Lockheed Martin that only 80 percent of aircraft at a given site would be available for ETOS calculation, they mandated more than their proposed four aircraft. They pointed to a subsequent EN response that addressed only the necessary five aircraft for low-rate initial production, not for ETOS calculation. In this technical area, GAO found for the Navy's assessment.

The Navy assessed past performance by asking offerors to provide related contract experience and to describe performance in meeting technical and quality requirements. Of particular importance to the evaluation was that when problems were encountered, the offeror was expected to note *systemic improvement* efforts applied to those problems and the impact of the efforts in resolving the problems and to provide records of those results that had been documented in government record systems. On balance, the past performance of General Atomics on Predator or derivative aircraft was held by the Navy to be marginal. In Air Force contract performance assessment reports on MQ-9 Reaper, a Predator derivative aircraft, General Atomics was noted to have experienced issues with systems engineering, software engineering, and contract execution due to schedule issues. In Army performance assessments on MQ-1C ER/MP, another Predator derivative aircraft system, it was noted that performance was marginal for quality, schedule, and cost performance. Furthermore, in Army assessments on I-GNAT, yet another Predator derivative, and despite contrasting scores from Army and Defense Contract Management Agency representatives, the record of General Atomics' poor performance on quality, schedule, and cost was more credible. In those three cases, GAO found that the Navy's evaluation was reasonable. Overall, GAO determined that Lockheed Martin's challenges to the evaluation of its team's past performance provided no basis for questioning the Navy's determination that the Lockheed Martin team—in particular, General Atomics—had a recent poor past performance.

By contrast, GAO supported the Navy's finding that the Northrop Grumman division serving as the prime contractor and system integrator— Northrop Grumman integrated systems Eastern region (Bethpage)—had satisfactory to good performance, demonstrating "high quality technical performance" on each of seven relevant contracts and favorable or excellent cost control on six of those contracts. Furthermore, GAO supported the Navy's finding that the overall Northrop Grumman team demonstrated excellent program management on most relevant contracts. However, GAO noted the Navy's concern with the Northrop Grumman team's aircraft provider, its integrated systems Western region (Rancho Bernardo), with regard to its overall rating as marginal. The office supported the Navy's determination that a record of systemic improvement on evaluated contracts had been demonstrated and that the position of the Eastern region (Bethpage) as prime for the team mitigated the concerns to a large degree.

One month after GAO announced it was sustaining Boeing's KC-X protest, and one hundred days since the Lockheed Martin BAMS protest, GAO ruled in favor of the Navy's award to Northrop Grumman on BAMS, denying the Lockheed Martin/General Atomics protest. The ruling could not have been received at Northrop Grumman with greater relief or greater satisfaction. In summary, GAO found that the Navy had

> reasonably determined that the awardee (Northrop Grumman) had significant advantage over protester with respect to past performance where: protesters' subcontractor, responsible for approximately 50 percent of contract effort, had recent past performance history of being unable to resolve staffing and resource issues, resulting in

adverse cost and schedule performance on very relevant contracts for unmanned aircraft; the record did not demonstrate that the protester's subcontractor had implemented systemic improvement that resulted in improved performance; while the operating division of awardee (Northrop Grumman) also had performance problems on very relevant contracts for unmanned aircraft, many had been addressed through systemic improvement; and, overall performance of the awardee's team on most evaluated contract efforts was rated better than satisfactory, while overall performance of protester's team on 11 of 26 contract efforts was only marginal.[4]

The Navy immediately restarted the program, and Northrop Grumman set off on preparing for a preliminary design review. Lockheed Martin then had further discussions with the Navy that suggested they were strongly considering taking their case to the federal court. Several conference calls ensued with Lockheed Martin and NAVAIR leadership, wherein the Navy convinced them that continuing down this path with their protest would be futile, nor would it help in their future business dealings with NAVAIR. As a result, Lockheed Martin dropped all further protest efforts.

Within the next few weeks, Northrop Grumman leadership received several inquiries from Lockheed Martin executives inquiring if there might be any role for them on the winning BAMS team once all this had been resolved. Lockheed Martin's Eagan site was now in serious trouble from a business perspective, and they were looking for any and all ways to save the business. After the bruising competition and friction that occurred from the protest, the Northrop Grumman team had no place for a competitor. The central role of General Atomics on their team and in the protest, and the Navy's low opinion of that team as a result, meant that there was nothing Lockheed Martin could add that would better the Northrop Grumman team's position with the customer. Furthermore, every aspect of the supplier base through development and production was covered for the program. The Northrop Grumman BAMS team set off on the long journey to actually develop the BAMS system and ultimately the Northrop Grumman MQ-4C Triton to provide real-time ISR over vast ocean and coastal regions for the U.S. Navy around the globe.

EPILOGUE

Leadership that creates a climate of effective performance is a critical ingredient in the quest to make strategy work. . . . Leadership is the vital force needed to create an execution-based culture in the organization.

—LAWRENCE G. HREBINIAK[1]

In his long-range guidance for the U.S Navy released in December 2018, CNO Adm. John Richardson called for rapidly acquiring key platforms to achieve high-velocity outcomes; among those key platform goals was reaching "MQ-4C Triton Initial Operational Capability [IOC] in 2021."[2] For any observer of defense acquisition not familiar with the long, tortuous history of Triton, it is sobering to contemplate that it will have taken more than twenty years from the initial BAMS concept study to reach Triton's IOC. The original concept was fairly simple and straightforward: use the Air Force Global Hawk, with early production Block 10s already flying, as the basis for transitioning a fairly mature unmanned aircraft system into Navy service with an adapted and improved sensor payload optimized for maritime surveillance.

Opportunities were in fact presented for an earlier and more achievable IOC. In the early gestation of BAMS, just as the Navy was drastically reducing its future years funding in late 2003 for BAMS to permit full funding for the manned MMA, Northrop Grumman responded to the Navy cuts with an offer to rephase the BAMS acquisition. The resulting rapid response proposal would have been executable under the revised and much reduced funding profile and would have made possible holding to an IOC in 2008, only nine months later than originally planned. The compromise would have meant a reduction in aircraft from twelve to eight inside the future years defense planning window and the stand-up of two

227

main operating bases rather than three. The MFAS radar would not have been available, requiring the addition of maritime modes to the existing Raytheon radar hosted in the early version RQ-4 aircraft. The requirement to descend from altitude for positive identification of targets below weather had not emerged at that point, so the aircraft would not have featured the more robust wing and fuselage structure with the deicing system. By delaying, the Navy obtained Triton, an unquestionably more capable aircraft. But by extending the development over ten more years, the Navy paid a premium for it and potentially had a gap of significant operational capability it could have used during Operation Enduring Freedom.

However, the most recent delay of IOC to 2021 was at least partially driven by a realignment of the program's acquisition strategy to accommodate development and fielding of a more robust sensor configuration. This followed instructions contained in a 2016 acquisition decision memorandum that directed the Navy to lay in funding for the robust sensor software upgrade. That configuration will provide for a signals intelligence capability and includes sensors, supporting software and hardware, and changes to permit processing of more highly classified information. The Navy intends for this more robust sensor configuration to replace the EP-3 Aries II manned aircraft for most missions.[3] Ironically, the original BAMS concept study by Northrop Grumman proposed a BAMS signals intelligence variant that now, at least twenty years after that study, might become an operational reality.

The requirement to descend from altitude during a surveillance mission emerged sometime between the PUMAS study and the ramp-up to the BAMS proposal effort. It was unexpected, and the initial judgment in Northrop Grumman was that a competitor had successfully shaped a BAMS requirement deliberately to disadvantage the Global Hawk aircraft's mission endurance at high altitude. It soon became apparent that this requirement was a cultural holdover from legacy maritime patrol and reconnaissance operations wherein P-3s routinely descended to lower altitude to obtain eyes on target. In internal discussions, Northrop Grumman operations analysts and systems engineers conceded that there would be some percentage of mission time that had to be flown lower to get below clouds. The question was whether the Global Hawk air vehicle structure could hold up over repeated descents and climbs over water without performance being further degraded by icing at lower and medium altitudes. In addition, turbulence and G loading could be more significant over water in the more difficult weather and turbulence conditions around the world in maritime environments.

During the capture phase, the Northrop Grumman team became aware of concerns by NAVAIR, particularly in the air vehicle structures competency, that Global Hawk could not structurally meet these climb and descent requirements. These concerns were due in part to both an inaccurate internal engineering memo written by Northrop Grumman and negative shaping by competitors. Meetings were immediately set up with NAVAIR engineering senior civilian Rich Gilpin and his engineers—an example of the Northrop Grumman BAMS campaign leader moving quickly to counter misperceptions by the customer and negative shaping by the competitors. Once again, strict adherence to the original win strategy theme regarding structural integrity and adequate growth margin became a signature selling point in the Global Hawk BAMS offering: the Global Hawk could accommodate future upgrades because it was not maxed out at the start; with almost four hundred pounds of weight margin, large unused wing and fuselage surface areas, and ample electrical power, changes to meet the "dip" could be accommodated without compromising significant mission performance.

As the first of several Tritons delivered to the Navy finally began regular operations, one could barely distinguish any external differences—other than exterior paint—between the Air Force Global Hawk and the Navy Triton. So why did this acquisition take such an inordinately long time from concept to fulfillment? Other successful transitions of aircraft models from one service to another have been more noteworthy for their relatively smooth execution. The U.S. Navy A-7 attack aircraft, a complex design selected by Navy contract in 1964 and flown in 1965, was that year directed to enter Air Force inventory by the secretary of defense. The Air Force adopted the A-7 but with significant changes to the Navy configuration that included an improved engine, a new avionics system with heads-up display, rotary gun system, and boom receptacle refueling system. Two squadrons of Air Force A-7Ds deployed for combat operations just seven years later. Granted, the urgency of combat operations in Southeast Asia likely accelerated the transition, but when one considers that Navy adoption of an unmanned Air Force aircraft took more than twenty years from concept to initial operational capability, other factors unique to this transition must have had undue influence on the acquisition process.

That Triton exists at all is a testament to the emergence of innovative and single-minded leaders who in every instance defied the norm or the conventional wisdom to impose their own rules—leaders with the vision to formulate a winning strategy and the confidence and discipline to ensure its implementation. That prescription is what elevated the BAMS concept to its acceptance as a defined Navy requirement and, with new leaders and new strategy formulated for a new set of programmatic conditions, forged the winning proposal that could succeed against a field of highly competitive industry giants.

Some would say leadership is an intangible force that makes itself known by the influence it exerts, often recognizable only in retrospect by calculating the change it has effected. In the broadest sense, that may be true. In the BAMS-to-Triton saga, leadership was clearly on display at several points over twenty-plus years that could be distilled to a few consistent characteristics: an understanding of the Navy customer and business (the maritime patrol and reconnaissance mission); an openness to reimagining operational concepts; the vision to discern a value added by new technologies, with the experience and insight to glean compelling discriminators from those technologies; and the standing, comportment, persona, and persistence to articulate the vision and to compel others to adopt it. Strategy then follows.

The leaders described in this book possessed the characteristics that made them natural progenitors of well-formed strategies, which rested in some cases on their being very aggressive and setting unconventional conditions and tactics to outflank competitors. Disruptions and disrupters were resources, and often obstacles, that the leader had to harness in service to the strategy. Above all, the leaders had to show discipline in holding to the strategy when unknowns surfaced, but they also had to show adaptability in recognizing setbacks or changes and adjusting the strategy to counter or circumvent. In many ways, these highly complex competitive captures, where billions of dollars and the future of a company's growth are at stake, are much like combat, where simple but compelling strategies are devised to win the war and are then countered by competitors (the enemy) who make opposing moves. Offensive and defensive strategies are employed continually, there are setbacks and losses, but the capture team meets to devise its next move, with winning the contract as the ultimate objective. Competitive captures of this magnitude, either strategically or financially, can make or break a company. Winning is everything, and losing is incomprehensible. In addition, the fostering of this competitive environment involving military requirements and technology brings forward only the finest of weapons systems that are employed to defend our great nation.

Mention has been made of the Northrop Grumman business acquisition process, which offers guidance and discrete steps that look almost indistinguishable from business development processes employed by any large aerospace and defense company. Most companies begin their strategy development by identifying a value proposition that is well mapped to the customer's requirements, whether those requirements are explicitly detailed in a document or derived implicitly from discussions with buyers and operators. That value proposition begins to reveal areas of relative strength and competitive weakness that need to be addressed through investment, teaming, demonstration, or advertising. With the value proposition understood, companies formulate a win strategy that magnifies their strengths, neutralizes any perceived weaknesses, and targets competitor strengths with counter tactics. Understanding the customer's process for selection is paramount to aligning the win strategy with scoring, available funding, risk aversion, and timeline to press down on those win strategy tactics that cannot be compromised and to address those that require more emphasis or funding to mature.

Perhaps key to understanding the customer's process is divining what the ideal end state for the customer will be. A customer's end state is most often determined by the choice of individuals who have some degree of influence throughout the acquisition process rather than by large monolithic organizations. Knowing those individuals and their acquisition history, preferences, predilections, and preferred outcome informs how best to pre-sell and manage expectations: "The essence of a winning strategy lies in differentiating your solution from that of [the] competitors in the specific attributes deemed to be the most important to the key decision makers. These attributes, or customer values, once identified, guide the strategy formulation."[4]

The authors' experience is that a dogged adherence to a business acquisition process will enhance the probability of winning simply by forcing the capture team members through the steps that organize their efforts. The process is not a guarantee of winning, but not adhering to it could set up the company to miss a critical step that could cause it to lose the competition. Yet it remains the role of the leader to bring to bear his or her vision, to ensure the capture team owns the win strategy, and to drive the capture team to execute the tactics that undergird the win strategy. The leader gives impetus to the process and motivates the capture team to fulfill a vision that he or she has declared. At two distinct pivotal moments in the long Triton history, leaders came to the fore with a vision, architected a strategy, and drove an outcome that was critical to the end state: from 2000 to 2002, at the inception of BAMS, and from 2006 to 2008, for a Northrop Grumman business capture then pitted against major industry competition.

To be sure, many others played a part. Those in more senior positions that encouraged and endorsed the campaign and capture leaders deserve credit. Throughout the history of BAMS, others in more senior positions in the company lacked the customer-centric orientation and cultural familiarity to perceive the customer's desired end state and how the Global Hawk was being positioned for it. More often, these senior people were too invested in intrasector rivalries and career advancement. The lesson here is that if the leader's vision cannot be imparted and embraced, others will become noncontributors at best and impediments to winning and business growth at worst. But what also is critical is having a vision for what a win of this magnitude could ultimately do for the business and the company. A push by Northrop Grumman into this new domain could forever change the future of the company. This kind of a vision was a ten- to twenty-year vision, and only a handful of leaders understood this.

"BEST VALUE" MUST BE MADE MEASURABLE

For BAMS or any other major competition, win strategy must reflect the leader's vision, must position the offering in the most advantageous light, and must relate to the customer's requirement in a way that it can be perceived as a value added. The BAMS win strategy, formulated in 2006 by the campaign lead and capture executive, mandated a nondevelopmental, low-risk approach requiring minimum change to the Air Force RQ-4B Global Hawk. Mindful of the origin of the BAMS concept as a Global Hawk derivative, this seemed a logical way to ensure that new, high-risk enhancements did not creep into the basic offering but that where the intended Navy Global Hawk derivative RQ-4N held a clear technical advantage over the competitors, Northrop Grumman was accorded full credit in a best-value runoff. To hold to that logic, the win strategy necessarily also mandated technology readiness level six or better for holding risk in abeyance; and to create daylight between the RQ-4N and the competitive field and draw out the inherent advantage of the RQ-4N platform for the BAMS mission, the win strategy made a fundamental goal of meeting all threshold requirements and greater than 90 percent of objective requirements. Providing a differentiated and increased value to the Navy customer was always a focus of the team.

As the Navy's acquisition machinery began to take hold of requirements first expressed in the November 2006 draft capability development document and either tightened or loosened them to preserve the appearance of a level competitive playing field through the draft RFP, the importance attached to key technical elements of the win strategy began to come into clear focus. Late 2006 through 2007 presented opportunities to engage with the customer by means of formal letters to the principal contracting officer in an attempt to tighten the obvious drift in technical requirements. For example, the draft RFP contained no objective requirement with regard to ETOS that, in the original capability development document, had mandated 95 percent ETOS at 3,000 nautical miles. An objective requirement strengthened the relative position of the RQ-4N when judged by best value. The letter to the contracting officer yielded a restoration of the ETOS objective in the final RFP. Similarly, in the draft RFP, sensor payload was reduced to simple thresholds, with objectives reduced or made ambiguous. Two strongly worded letters to NAVAIR resulted in more thresholds and objectives restored to the final RFP, with added radar modes and resolution and range requirements that reinforced the selection of MFAS as the primary sensor.

The draft RFP dropped any objective requirements for time-to-station. In effect, this penalized the RQ-4N by not giving credit for its speed advantage over competitor air vehicles. Again, a letter to NAVAIR argued for restoration of the original objective requirement of ten hours' time-to-station, which Northrop Grumman could not only meet but also do with the confidence to guarantee it. The shaping of the technical requirements in the final RFP to criteria more favorable to best-value comparison was due to the efforts of the Northrop Grumman capture campaign in their engagements with the customer, based on the working assumptions of the win strategy. This strategy ultimately proved to be brilliant. There was a relentless effort to dialogue with the customer and not give up on these requirements, because this would only dilute the critical requirements and potentially level the playing field for the competition. This kind of tenacious, driven, and focused capture strategy is what separates the winner from the losers. In the end, the team that wants it the most is usually the team that wins; this was certainly the case for Northrop Grumman, which spent more than ten years pursuing this procurement.

But the march to making Triton a viable operational ISR resource clearly did not end with winning the BAMS competition in 2008. Further years of development were required, beginning with formal acquisition process steps immediately after the contract award that were kicked off by an acquisition decision memorandum issued in April 2008. With that memorandum, BAMS entered a system development and demonstration phase that aimed toward a critical design review in 2011—all necessary steps called out in the Pentagon's major systems acquisition framework. A key consideration in the critical design review was change imposed on the basic Air Force Global Hawk to navalize the aircraft and its systems. During the course of the competition, the Navy evolved an operational requirement for BAMS to fly to lower altitudes to get below the weather for positive identification of surface targets. This necessitated introduction of an anti-icing system, anti-lightning protection, and a more robust wing and fuselage structure.

The greatest change from the Air Force model was in the sensor package. In early 2012 flight-testing began on the BAMS multifunction active sensor (MFAS) integrated on the company's Gulfstream GII testbed aircraft at Palmdale, California. In 2007 MFAS had been hotly debated in the formulation of the BAMS win strategy, with some in the Northrop Grumman unmanned systems division believing it to be too great a departure from the basic Air Force configuration. The decision to include MFAS in the BAMS sensor package offering was not endorsed by all in Northrop Grumman, but leadership vision and persistence prevailed, and the Navy evaluators were delighted with the system, which "operates with a rotating sensor that incorporates electronic scanning and provides mode agility to switch between various surveillance methods. These include maritime-surface-search mode for tracking maritime targets and inverse-synthetic-aperture radar mode for classifying ships. Two synthetic aperture radar (SAR) modes are used for ground searches; spot SAR for images of the ground and stationary targets and strip SAR for images along a fixed line."[5] The sensor was a clear departure from the Air Force Global Hawk and was a response to the Navy's unique maritime operating environment.

In offering MFAS, Northrop Grumman again closely adhered to its evolved win strategy in providing state-of-the-art sensors despite the additional cost to their proposal. The win strategy themes keyed on offering a modern 360-degree AESA and on offering no system at less than technology readiness level six. MFAS, tested in the company-funded Head Start risk reduction effort, could be assessed at technology readiness level seven based on test results, and it provided objective-level performance at delivery, with future growth capability that would not be maxed out at the moment operational capability was declared.

Northrop Grumman's offering, given the derivative alphanumeric designation RQ-4N as a spin-off of the Air Force RQ-4B, received the Navy's official designation of MQ-4C shortly after the system development and demonstration contract award. With the unveiling of the first MQ-4C at Palmdale, California, in June 2012, the aircraft was christened officially with its nom de guerre, Triton. Its first flight preceded the unveiling by mere days. An eighty-minute flight to 20,000 feet from the company's location in Palmdale marked the start of a lengthy flight-testing schedule. Rear Adm. Mat Winter, the Navy program executive officer for unmanned aviation and strike weapons, commented on the occasion: "This flight represents a significant milestone for the Triton team. The work they have done and will continue to do is critical to the future of naval aviation, particularly to our maritime patrol and reconnaissance community."[6]

If there was a point at which Triton clearly turned a corner and began to make great programmatic strides, it was 2016. In the second quarter of the fiscal year, Triton completed its operational assessment, during which a combined Navy and Northrop Grumman integrated test team validated the aircraft's sensor performance at different altitudes and ranges. Triton's "ability to classify targets and disseminate critical data"[7] was shown, and on-station endurance performance was confirmed. Completion of the operational assessment was a prerequisite for entering a milestone C review in September that conferred approval for low-rate initial production. Exiting the review successfully, the Navy ordered a first production lot of three MQ-4Cs plus main and forward operating control stations.[8]

In July 2018 Unmanned Patrol Squadron (VUP-19) successfully conducted its first flights as a squadron operating the MQ-4C Triton. Known by the nickname "Big Red," the squadron flew both of its MQ-4Cs, which Northrop Grumman had previously delivered to Naval Base Ventura County, Point Mugu, California. These flights marked the realization of an operational capability after many years of development and the preparation of the squadron for operational deployment. The Tritons took off from California operated by pilots from both VUP-19 and the Navy's Air Test and Evaluation Squadron One (VX-1) under local line-of-sight control. Pilots controlled flights from a forward operating base as part of the squadron's detachment, sited in Point Mugu to conduct the launch, recovery, and maintenance of the MQ-4C. Following takeoff under the control of the local launch station in Point Mugu, control of the MQ-4C was passed to the main operating base at Naval Air Station Jacksonville, Florida, where VUP-19 and VX-1 pilots, flight officers, and sensor operators maintained operational control of aircraft. For landing following the flight, control was passed back to the forward operating base pilots in California. VUP-19 will continue operating Tritons out of Point Mugu as they prepare for the aircraft's eventual move to Andersen Air Force Base in Guam. From there, Tritons will establish an early operational capability in support of the U.S. Seventh Fleet.[9]

Finally, in surveying the long acquisition history of the MQ-4C, what becomes abundantly clear is that factors wholly unique to the BAMS acquisition and the BAMS customer existed that demanded a highly specialized and tailored approach from the BAMS industry capture campaign team at several junctures throughout that history. On many large, complex acquisitions—and BAMS fit squarely in that category—the customer often does not have a firmly grounded understanding of his or her requirements. For the Navy BAMS customer, undertaking the service's first large unmanned aircraft system acquisition, this was the case, and once the customer understood that requirements could not be set in isolation from the aircraft system experts, it presented an opening to shape the dialogue with the customer on what was important in assessing UAV candidates.

What the BAMS experience reveals, however, is that the company with the greatest name recognition, deepest pockets for discretionary investment, or the most innovative technical solution is not guaranteed to be the eventual selectee. Rather, the competitor that follows a prescribed and tested business acquisition process and refines a compelling win strategy that builds on some preexisting market advantage or position will create the best circumstance for winning. But again, visionary and tenacious leaders who are aggressive and relentless also play a significant qualitative role in winning. Northrop Grumman rightly perceived that it possessed a distinct advantage going into the proposal phase at the point of the concept study contract in 1999, at the first BAMS competition that faltered from lack of government commitment in 2004, and in the decisive proposal effort that kicked off in 2006 and brought a competitive win finally in 2008.

But that advantage going in is always at risk of being nullified by a lack of executive leadership and commitment, a lack of discipline in executing the win strategy, a failure to relate one's own team's strengths to the customer's most desired attributes (and failure to correct own-team deficiencies, or least reduce perceived risk in the customer's judgment), acting on poor intelligence with regard to the customer's funding profile and available budget and setting a price outside the competitive range, an inability to adapt the strategy to changing circumstances, a failure to discern and counter threatening competitor actions, and finally, standing back when the proposal is submitted to await the outcome of the evaluation. In this story, the BAMS capture team did not stand down during the post-proposal phase but used the time when the Navy evaluation was ongoing to continue to shape the evaluation through the EN process. Entering the post-proposal submission period, the BAMS capture campaign team was fortunately alerted by the aggressive and ultimately successful protest of the declared Northrop Grumman win on KC-X and acted on that warning shot to devote equal attention to the protest strategy and its execution.

Stepping back from the more unique aspects of the BAMS acquisition, there are some overarching lessons that companies can learn, or relearn, when deciding whether to pursue an opportunity. All were highlighted to some degree in the preceding pages and mentioned when they had an impact on the BAMS acquisition. They apply to the BAMS capture, but equally to any large, complex military system competition:

- Approach every capture as a combat situation.
- Keep the win strategy simple and compelling.
- Shape requirements and defend against competitive shaping.
- Team internally and externally only for strength.
- Don't lose sight of the fact that the customer has no tolerance for cost growth.
- Poor past performance cannot be disguised and must be mitigated in some way.
- Understand "cost as an independent variable" (CAIV), estimates and government most probable cost.
- Best value can win, and sometimes lowest cost can; however, the best technical offering and lowest cost almost always wins.
- Expect a protest.

NOTES

INTRODUCTION

1. Jonathan Hoppe, "Survey Results: What's the Greatest Naval Innovation and Why?" *USNI News*, May 10, 2016.
2. Craig C. Felker, *Testing American Sea Power* (College Station: Texas A&M University Press, 2007), 2.
3. William F. Trimble, *Admiral William A. Moffitt, Architect of Naval Aviation* (Annapolis, MD: Naval Institute Press, 1994), 71.

CHAPTER 1. INCEPTION

1. "P-8 Introduced at Roll Out Ceremony," *Planeside*, no. 1 (2012), https://maritimepatrolassociation.org/wp-content/uploads/2019/02/MPA_Newsletter_2012_1_compress.pdf.
2. "Commissioning Pennant," Naval History and Heritage Command, May 24, 2017, https://www.history.navy.mil/browse-by-topic/heritage/customs-and-traditions0/commissioning-pennant.html.
3. Northrop Grumman, "Northrop Grumman Delivers First Operational MQ-4C," November 10, 2017, https://news.northropgrumman.com/news/releases/northrop-grumman-delivers-first-operational-mq-4c-triton-to-us-navy.
4. Todd Miller, "U.S. Navy MQ-4C Triton Makes Persistent Progress Towards Deployment," *Second Line of Defense*, March 1, 2016, https://sldinfo.com/2016/03/us-navy-mq-4c-triton-makes-persistent-progress-towards-deployment.
5. Department of the Navy, Office of Budget, *Highlights of the Department of the Navy FY2015 Budget* (2014), 4–12, http://www.finance.hq.navy.mil/fmb/15pres/Highlights_book.pdf.
6. PEO (U&W) Public Affairs, "MQ-4C Triton UAS Demonstrates New Capabilities During Flight Test," *NAVAIR News*, June 22, 2016.
7. Tyler Rogoway, "The Navy Has the Ultimate MH370 Search Tool, It's Just Not Operational," *Foxtrotalpha.com*, March 18, 2014, https://foxtrotalpha.jalopnik.com/why-mq-4c-triton-the-ultimate-mh370-search-tool-isnt-1545912657.
8. Joint Requirements Oversight Council, *Mission Need Statement for Long Endurance, Reconnaissance, Surveillance, and Target Acquisition Capability* (JROC MNS 003-90), January 5, 1990, https://fas.org/irp/doddir/usaf/conops_uav/part01.htm.
9. Rick Thomas, *Global Hawk: The Story and Shadow of America's Controversial Drone* (self-published, December 2015), 73.

10. Bradd C. Hayes and Douglas V. Smith, "The Politics of Naval Innovation," Center for Naval Warfare Studies Report 4-94 (Newport, RI: U.S. Naval War College, August 1, 1994), 79.

11. Hayes and Smith, 10.

12. Hayes and Smith, 79.

13. Hayes and Smith, 87.

14. Andrew F. Krepinevich Jr., *The Military-Technical Revolution: A Preliminary Assessment* (Washington, DC: Center for Strategic and Budgetary Assessments, October 2, 2002), 52, https://csbaonline.org/research/publications/the-military-technical-revolution-a-preliminary-assessment.

CHAPTER 2. DISRUPTERS TO THE FORE

1. Jay Samit, *Disrupt You! Master Personal Transformation, Seize Opportunity, and Thrive in the Era of Endless Innovation* (New York: Macmillan, 2015), 21.

2. Russell Schuhart, "Prepare for the 'To Be or To Do' Moment," U.S. Naval Institute *Proceedings* 143, no. 12 (December 2017): 11.

3. Eduardo P. Braun, *People First Leadership* (New York: McGraw-Hill, 2017).

4. Thomas Kilcline, "Fitness Report and Counseling Record on LCDR Mark Turner," BUPERS 1610-1, June 7, 2002, personal record provided by Turner.

5. Peter Swartz, "Organizing OPNAV (1970–2009)," Center for Naval Analyses Study (Alexandria, VA: Center for Naval Analyses, January 2010), 58.

6. Swartz, 68.

7. George Spangenberg, oral history transcript, August 31, 1997, 300, https://www.docdroid.com/x9czNLE/george-spangenberg-oral-history-pdf#page=3.

8. Thomas, *Global Hawk*, 47.

9. Thomas, 60.

10. Rebecca Grant, "Eyes Wide Open," *Air Force Magazine* (November 2003): 39.

11. Thomas, *Global Hawk*, 62.

CHAPTER 3. TURBULENCE

1. John Mintz, "Aspin Sees Military Promoting Clinton Industrial Policy," *Washington Post*, September 2, 1993.

2. Dan Balz and David Broder, "President Clinton's First 100 Days," *Washington Post*, April 29, 1993.

3. Steven V. Roberts, "Congress: The Provocative Saga of the $400 Hammer," *New York Times*, June 13, 1984.

4. Robert S. Leonard and Jeffrey A. Drezner, *Global Hawk and Dark Star* (Santa Monica, CA: RAND, 2002), 2.

5. Jeffrey Drezner, Geoffrey Sommer, and Robert S. Leonard, *Innovative Management in the DARPA HAE UAV Program* (Santa Monica, CA: RAND, August 24, 2020), xviii.

6. Richard Van Atta, *Transformation and Transition: DARPA's Role in Fostering an Emerging Revolution in Military Affairs* (Alexandria, VA: Institute for Defense Analyses, April 2003), 47.

7. Thomas, *Global Hawk*, 64.

8. Thomas, 68.

9. Thomas, 71.

10. Geoffrey Sommer, *The Global Hawk Unmanned Aerial Vehicle Acquisition Process* (Santa Monica, CA: RAND, 1997), 4.

11. Leonard and Drezner, *Global Hawk*, 12.

12. Thomas, *Global Hawk*, 83.

13. Thomas, 72–76.

14. Thomas, 80.

15. Bill Kinzig, *Global Hawk System Engineering Case Study* (Wright-Patterson Air Force Base, OH: Air Force Center for Systems Engineering, 2010), 35.

16. Kinzig, 36.

17. Thomas, *Global Hawk*, 87.

18. Thomas, 100.

19. Thomas.

20. John Deutch, "Consolidation of the U.S. Defense Industrial Base," *Acquisition Review Quarterly* (Fall 2001).

21. Justin Wakefield, "A Second Helping of the Last Supper," *Journal of Contract Management* (Summer 2014).

22. Richard Burnett, "Grumman Could Be Florida-Bound," *Orlando Sentinel*, February 5, 1990.

23. Kenneth Gilpin, "Takeover Tables Turning on Northrop, Some Contend," *New York Times*, March 17, 1994.

24. Eileen Hogan, *The HR Side of Northrop Grumman's Acquisition Process* (Alexandria, VA: Society for Human Resource Management, 2009), 6.

25. Hogan, 29.

26. "LTV Puts Aerospace-Defense Unit Up for Sale," *Defense Daily*, May 21, 1991.

27. "Vought Aircraft Industries, Inc.," *International Directory of Company Histories*, vol. 49 (St. James, MO: St. James Press, 2003).

28. Darcy Jacobsen, "6 Big Mergers That Were Killed by Culture," *WorkHuman.com*, September 26, 2012, https://www.workhuman.com/resources/globoforce-blog/6-big-mergers-that-were-killed-by-culture-and-how-to-stop-it-from-killing-yours.

29. Chris Barrett et al., "Breaking the Culture Barrier in Postmerger Integrations," Boston Consulting Group, January 13, 2016, https://www.bcg.com/publications/2016/breaking-the-culture-barrier-in-postmerger-integrations.

CHAPTER 4. MOMENTUM

1. General Colin Powell, Bottom-Up Review press briefing, September 1, 1993, https://www.c-span.org/video/?49768-1/defense-department-review.

2. Mark Bowden, *Black Hawk Down—A Story of Modern War* (New York: Grove Press, 1999), 112–13.

3. Norman Polmar and Kenneth J. Moore, *Cold War Submarines: The Design and Construction of U.S. and Soviet Submarines* (Washington, DC: Potomac Books, 2004), 331.

4. Martin Ferber, "Tactical Aircraft: Issues Concerning the Navy's Maritime Patrol Aircraft," General Accounting Office Report to the House Armed Services Committee, NSIAD-91-229 (Washington, DC: U.S. Government Printing Office, September 4, 1991), 9.

5. David Reade, "New Developments," *Maritime Patrol Aviation* 2, no. 4 (March 1993): 58.

6. Reade, 48.

7. Gary T. Ambrose, "Transforming Maritime Patrol Aviation" (Quantico, VA: U.S. Marine Corps Command and Staff College, 2003), 12–13.

8. John T. Correll, "The Legacy of the Bottom-Up Review," *Air Force Magazine* (October 2003): 54.

9. Correll, 55.

10. Correll.

11. Correll, 59.

12. N. D. Phan, "P-3 Service Life Assessment Program (SLAP)—A Holistic Approach to Inventory Sustainment for Legacy Aircraft," in *Proceedings from Tri-Service Corrosion Conference* (Patuxent River, MD: Naval Air Systems Command, 2003), 1–13, http://namis.alionscience.com/conf/tscc/search/pdf/ AM025612.pdf.

13. Daniel Dolan, "When ASW Didn't Matter," *Proceedings* 143, no. 8 (August 2017): 62.

14. Dolan.

15. Thomas, *Global Hawk*, 106.

16. Vago Muradian, "Northrop Grumman to Buy UAV-Maker Ryan for $140 Million," *Defense Daily,* May 28, 1999.

17. Thomas, *Global Hawk*, 106.

18. Swartz, "Organizing OPNAV."

CHAPTER 5. DISRUPTION

1. Naval Studies Board of the National Academy of Science. *Autonomous Vehicles in Support of Naval Operations* (Washington, DC: National Academies Press, 2005), 106.

2. Clayton Christensen, Michael Raynor, and Rory McDonald, "What Is Disruptive Innovation," *HBR's 10 Must Reads* (Boston: HBR Press, 2017), 3.

3. John Walker, "Why Innovation Must Go Beyond Disruption," *WIRED*, February 2015, https://www.wired.com/insights/2015/02/why-innovation-must-go-beyond -disruption/.

4. Alan Easterling, "MMA-BAMS Relationship," background paper in personal collection, September 1, 2004.

5. "Navy Extends Deadline to May 12 for BAMS Proposals," *Defense Daily*, May 4, 2000.

6. Easterling, "MMA-BAMS Relationship."

7. Jay Johnson, "You Are Now on the Cutting Edge," address, Current Strategy Forum (Newport, RI: U.S. Naval War College, June 15, 1999).

CHAPTER 6. GAINING A FOOTHOLD

1. Michael O'Guin, "Shaping the Battle Before It Begins," *APMP Journal* (Fall 2013).

2. Thomas, *Global Hawk*, 118.

3. Thomas, 115.

4. Easterling.

5. Maris Lapins, "Shaping the BAMS Requirement," private papers, May 5, 2004.

6. Easterling, "MMA-BAMS Relationship."

7. Easterling.

8. Easterling.

9. Stephen Trimble, "RQ-4N Spreads Global Hawk Brand to Maritime Patrol," *Flight International*, February 26, 2009, https://www.flightglobal.com/news/articles/rq-4n -spreads-global-hawk-brand-to-maritime-patrol-323093/.

10. Easterling, "MMA-BAMS Relationship."

11. Marc Strauss, "Navy Outlines Future UAV Strategy," *Defense Daily*, February 20, 2002.

12. Lapins, "Shaping."

13. Lapins.

14. Lapins.

15. Brendan Rivers, "UAV Test Beds Ordered by U.S. Navy," *Journal of Electronic Defense*, March 1, 2003.

16. B. C. Kessner, "Predator B Variant to Challenge Global Hawk for Navy BAMS UAV," *C4I News*, May 1, 2003.

17. Kinzig, *Global Hawk*, 69.

18. Lorenzo Cortes, "Lockheed Martin and General Atomics Team Up," *C4I News*, September 4, 2003.

19. Cortes.

20. Cortes.

21. Elizabeth Bone, "Unmanned Aerial Vehicles: Background and Issues for Congress" (Washington, DC: Congressional Research Service, Report RL31872, April 25, 2003), 41.

22. Irv Blickstein et al., *Navy Planning, Programming, Budgeting, and Execution—A Reference Guide for Senior Leaders, Managers, and Action Officers* (Santa Monica, CA: RAND, 2016), 5.

CHAPTER 7. INTERREGNUM

1. Captain Dennis Sorensen, USN, "Global Hawk UAV Makes First Flight for the Navy," Program Executive Office for Unmanned Systems and Weapons, Press Release, October 7, 2004, https://www.navair.navy.mil/node/13041.

2. Department of the Air Force, RDT&E Project Justification, Exhibit R-2, "PE#0305220F Global Hawk Development/Fielding," February 2005.

3. Department of the Navy, RDT&E Project Justification, Exhibit R-2a, "PE#0605500N Multi-Mission Maritime Aircraft," February 2004, 138.

4. Maris Lapins, "Revised BAMS RRP—Issues and Summary of Changes," *Program Notes*, author's personal collection, December 18, 2003.

5. Maris Lapins, "Notes on PBD Timeline," *Program Notes*, author's personal collection, January 2004.

6. Thomas, *Global Hawk*, 164.

7. Amy Butler, "Sidestepping Sole-Source to Northrop Grumman, Navy to Compete BAMS Work," *Defense Daily*, March 17, 2004.

8. Kinzig, *Global Hawk*, 73.

9. Alan Easterling, "Notes on P-8 MMA," *Program Notes*, author's personal collection, January 2004.

10. Easterling, "MMA-BAMS Relationship."

11. Richard Burgess, "Navy Moves Toward Goal of More Surveillance with Fewer Aircraft," *Sea Power*, June 1, 2004.

12. Easterling, "Notes on P-8 MMA."

13. Maris Lapins, interview by Edgar Snowden, November 19, 2017.

14. Lawrence G. Hrebiniak, *Making Strategy Work* (Upper Saddle River, NJ: FT Press, 2013), 295.

15. Naval Studies Board, *Autonomous Vehicles in Support of Naval Operations*, 3.

16. Patrick G. Roche, "A FORCEnet Framework for Analysis of Existing Naval C4I Architectures," master's thesis, Naval Postgraduate School, June 2003, 1.

17. Alan Easterling, interview by Ernest Snowden, April 28, 2017.

18. Easterling.

19. Howard Frauenberger, "Notes on PUMAS Study," *Program Notes*, author's personal collection, July 2006.

20. Frauenberger.

21. Frauenberger.

22. Easterling, interview.

23. Frauenberger, "Notes on PUMAS Study."

24. Captain Paul S. Morgan, "Performance under Contract N00019-05-C-0064 PUMAS," Department of the Navy Letter 1600, Serial PMA-263/063, May 9, 2006.

25. Sharon Anderson, "U.S. Second Fleet—The Fleet Lead in Joint Expeditionary Force Experiment 2006," *CHIPS*, July 1, 2006.

26. Anderson.

27. "News Breaks," *Aviation Week and Space Technology*, August 21, 2006, 27.

28. "News Breaks," *Aviation Week and Space Technology*, December 18, 2006, 10.

29. Department of the Navy, Navy Future Years Defense Plan, Exhibit R-2, "RDT&E Budget Item Justification (BA-7)," February 2007.

CHAPTER 8. SHAPING THE CAPTURE STRATEGY

1. David Bond, "Low Profile," *Aviation Week & Space Technology*, February 26, 2007, 23.

2. The definition or differentiation between the terms *capture* and *campaign* was vigorously debated at the Northrop Grumman corporate level with no consensus resolution. For the purpose of the BAMS narrative, the team settled on definitions that were generally accepted going forward. A capture is the pursuit (with the goal of winning) of a specific government or Department of Defense procurement for a product or system requirement, where the customer has, or will be, releasing a request for proposals. A campaign refers to the overarching effort surrounding the capture that involves the

people, places, and things that need to be organized and directed to win: the capture strategy, customer contacts, shaping or influencing the government's requirements, assessing the competition, moves and countermoves with the competition, the proposal development with multiple layers of color team reviews, and all the activities surrounding the successful winning of the procurement.

3. The integrated systems sector later became the aerospace systems sector, after Northrop Grumman leadership combined integrated systems with the space technology sector in January 2009.

4. Because of the number of contractor protests that were occurring in 2000–2008, the military services (and particularly the Navy) had become overly conservative on controlling information made available to contractors during the request for proposal phase. This actually caused greater confusion regarding the real requirements and the acceptable technical approach, and, for the industry competitors, determining exactly what the government customer wanted from the contractor. Shutting down an open dialogue between the government and contractors during request for proposal development was actually detrimental to providing cost-effective and top-quality solutions. The Northrop Grumman team began to fight this battle in the first meeting with Captain Morgan, the Navy's program manager for BAMS.

CHAPTER 9. COMPETING TO WIN

1. C. K. Pralahad, "Strategies for Growth," in *Rethinking the Future*, Rowan Gibson, ed. (Boston: Nicholas Brealey Publishing, 1998), 62.

CHAPTER 10. PROTEST

1. Stephen Trimble, "Lockheed Protests USN BAMS Award," *Flight Global*, May 5, 2008, https://www.flightglobal.com/lockheed-protests-usn-bams-award/80116.article.

2. Gary Kepplinger, "Decision in the Matter of Lockheed Martin MS2 Tactical Systems," U.S. General Accounting Office Report B-400135, August 8, 2008, 4, https://www.gao.gov/decisions/bidpro/400135.htm.

3. Elise M. Carlin, "GAO Sustains Boeing's Aerial Refueling Tanker Protest," *Federal Construction Contracting*, June 19, 2008, https://federalconstruction.phslegal.com/2008/06/articles/bid-protests/gao-sustains-boeings-aerial-refueling-tanker-protest-and-cites-significant-errors-in-the-procurement-process/.

4. Kepplinger, "Decision," 1.

EPILOGUE

1. Hrebiniak, *Making Strategy Work*, 25.

2. Admiral John Richardson, "A Design for Maintaining Maritime Superiority, Version 2.0," U.S. Naval Institute News, December 17, 2018, https://news.usni.org/2018/12/17/design-maintaining-maritime-superiority-2-0.

3. Director, Operational Test and Evaluation, "FY18 Navy Programs: MQ-4C Triton Unmanned Aircraft System," *FY2018 DOT&E Annual Report*, August 24, 2020, http://www.dote.osd.mil/pub/reports/FY2018/pdf/navy/2018mq4c_uas.pdf.

4. Michael O'Guin and Kim Kelly, *Winning the Big Ones—How Teams Capture Large Contracts* (self-published, 2012).

5. Jim Stratford, "U.S. Navy Sensor Flight Testing Begins and First MQ-4C Receives Its Wings," Northrop Grumman news release, February 27, 2012, https://news.northropgrumman.com/news/releases/photo-release-northrop-grumman-bams-unmanned-aircraft-system-program-achieves-two-major-milestones.

6. "MQ-4C Triton Broad Area Maritime Surveillance (BAMS)," *NavalDrones.com*, September 23, 2016, http://www.navaldrones.com/BAMS.html.

7. Northrop Grumman, "MQ-4C Triton Completes Critical Milestone to Expand Navy's Maritime Domain Awareness," September 26, 2016, https://news.northropgrumman.com/news/releases/mq-4c-triton-uas-completes-critical-milestone-to-expand-navys-maritime-domain-awareness.

8. Bill Carey, "U.S. Navy Orders First Three Production MQ-4C Tritons," *Aviation International News*, September 28, 2016, https://www.ainonline.com/aviation-news/defense/2016-09-28/us-navy-orders-first-three-production-mq-4c-tritons.

9. Matt Nemetz, "VUP-19 Conducts First MQ-4C Triton Flights," *Planeside* (3rd Quarter 2018): 15–16.

BIBLIOGRAPHY

Ambrose, Gary T. "Transforming Maritime Patrol Aviation." Quantico, VA: U.S. Marine Corps Command and Staff College, 2003.

Anderson, Sharon. "U.S. Second Fleet—The Fleet Lead in Joint Expeditionary Force Experiment 2006." *CHIPS*, July 1, 2006.

Balz, Dan, and David Broder. "President Clinton's First 100 Days." *Washington Post*, April 29, 1993.

Barrett, Chris, et al. "Breaking the Culture Barrier in Postmerger Integrations." Boston Consulting Group, January 13, 2016. https://www.bcg.com/publications/2016/breaking -the-culture-barrier-in-postmerger-integrations.

Bender, Bryan. "Navy Surveillance Plane Monitored Standoff in Bosnian Town." *Defense Daily*, October 10, 1996.

Blickstein, Irv, et al. *Navy Planning, Programming, Budgeting, and Execution—A Reference Guide for Senior Leaders, Managers and Action Officers*. Santa Monica, CA: RAND, 2016.

Bond, David. "Low Profile." *Aviation Week & Space Technology*. February 26, 2007.

Bone, Elizabeth. "Unmanned Aerial Vehicles: Background and Issues for Congress." Washington, DC: Congressional Research Service, Report RL31872, April 25, 2003.

Bowden, Mark. *Black Hawk Down—A Story of Modern War*. New York: Grove Press, 1999.

Braun, Eduardo P. *People First Leadership*. New York: McGraw-Hill, 2017.

Burgess, Richard. "Companies Studying Persistent Unmanned Maritime ISR for Navy." *Sea Power*, October 1, 2005.

———. "Navy Moves Toward Goal of More Surveillance with Fewer Aircraft." *Sea Power*, June 1, 2004.

Burnett, Richard. "Grumman Could Be Florida-Bound." *Orlando Sentinel*, February 5, 1990.

Butler, Amy. "Navy Proposes Delay to BAMS UAV Effort in FY 06 Budget." *Defense Daily*, September 17, 2004.

———. "Sidestepping Sole-Source to Northrop Grumman, Navy to Compete BAMS Work." *Defense Daily*, March 17, 2004.

Carey, Bill. "U.S. Navy Orders First Three Production MQ-4C Tritons." *Aviation International News*, September 28, 2016. https://www.ainonline.com/aviation-news/defense /2016-09-28/us-navy-orders-first-three-production-mq-4c-tritons.

Carlin, Elise M. "GAO Sustains Boeing's Aerial Refueling Tanker Protest." *Federal Construction Contracting*, June 19, 2008. https://federalconstruction.phslegal.com /2008/06/articles/bid-protests/gao-sustains-boeings-aerial-refueling-tanker-protest -and-cites-significant-errors-in-the-procurement-process/.

Christensen, Clayton, Michael Raynor, and Rory McDonald. "What Is Disruptive Innovation?" *HBR's 10 Must Reads*. Boston: HBR Press, 2017.

"Commissioning Pennant." Naval History and Heritage Command, May 24, 2017. https://www.history.navy.mil/browse-by-topic/heritage/customs-and-traditions0 /commissioning-pennant.html.

Correll, John T. "The Legacy of the Bottom-Up Review." *Air Force Magazine*, October 2003.

Cortes, Lorenzo. "Lockheed Martin and General Atomics Team Up." *C4I News*, September 4, 2003.

Department of Defense. "Navy Awards Northrop Grumman Unmanned Aircraft System Contract." Press release, April 22, 2008.

Department of the Air Force. RDT&E Project Justification, Exhibit R-2, "PE#0305220F Global Hawk Development/Fielding," February 2005. https://www.globalsecurity .org/military/library/budget/fy2013/usaf-peds/0305220f.pdf.

Department of the Navy. RDT&E Project Justification, Exhibit R-2a, "PE#0605500N Multi-Mission Maritime Aircraft," February 2004.

———. Navy Future Years Defense Plan, Exhibit R-2, "RDT&E Budget Item Justification (BA-7)," February 2007. https://www.secnav.navy.mil/fmc/fmb/Documents/08pres /rdten/RDTEN_BA7_BOOK.pdf.

Deputy Assistant Secretary of the Navy for Budget. *Highlights of the Department of the Navy FY 2021 Budget*. February 10, 2020. https://www.secnav.navy.mil/fmc/fmb /Documents/21pres/Highlights_book.pdf.

Deutch, John. "Consolidation of the U.S. Defense Industrial Base." *Acquisition Review Quarterly*, Fall 2001.

Director, Operational Test and Evaluation. "FY18 Navy Programs: MQ-4C Triton Unmanned Aircraft System." *FY2018 DOT&E Annual Report*, August 24, 2020. http:// www.dote.osd.mil/pub/reports/FY2018/pdf/navy/2018mq4c_uas.pdf.

Dolan, Daniel. "When ASW Didn't Matter." U.S. Naval Institute *Proceedings* 143, no. 8, August 2017.

Drezner, Jeffrey, Geoffrey Sommer, and Robert S. Leonard. *Innovative Management in the DARPA HAE UAV Program*. Santa Monica, CA: RAND, August 24, 2020.

Easterling, Alan. Interview by Ernest Snowden. April 28, 2017.

———. "MMA-BAMS Relationship." Author's personal papers. September 1, 2004.

———. "Notes on P-8 MMA." Author's personal papers. January 2004.

Fay, Matthew. "The Problem with Joint Aircraft Development." *Foreign Policy and Defense*, January 27, 2015.

Felker, Craig C. *Testing American Sea Power*. College Station: Texas A&M University Press, 2007.

Ferber, Martin. "Issues Concerning the Navy's Maritime Patrol Aircraft." General Accounting Office Report to the House Armed Services Committee, NSIAD-91-229. Washington, DC: U.S. Government Printing Office, September 4, 1991.

Frauenberger, Howard. "Notes on PUMAS Study." Author's personal papers. July 2006.

Fulghum, David. "Flying Wing, Four Others Selected for Tier 2+ UAV." *Aviation Week & Space Technology*, October 10, 1994.

Gardiner, Sam. "The Military-Technical Revolution." *RSAS Newsletter* 4, no. 3, August 1992.

Gilpin, Kenneth. "Takeover Tables Turning on Northrop, Some Contend." *New York Times*, March 17, 1994.

Grant, Rebecca. "Eyes Wide Open." *Air Force Magazine*, November 2003.

Hayes, Bradd C., and Douglas V. Smith. "The Politics of Naval Innovation." Center for Naval Warfare Studies Report 4-94. Newport, RI: U.S. Naval War College. August 1, 1994.

Hogan, Eileen. *The HR Side of Northrop Grumman's Acquisition Process*. Alexandria, VA: Society for Human Resource Management, 2009.

Hoppe, Jonathan. "Survey Results: What's the Greatest Naval Innovation and Why?" *USNI News*. May 10, 2016.

Hrebiniak, Lawrence G. *Making Strategy Work*. Upper Saddle River, NJ: FT Press, 2013.

Jacobsen, Darcy. "6 Big Mergers That Were Killed by Culture." *WorkHuman.com*, September 26, 2012. https://www.workhuman.com/resources/globoforce-blog/6-big -mergers-that-were-killed-by-culture-and-how-to-stop-it-from-killing-yours.

Johnson, Jay. "You Are Now on the Cutting Edge." Address, Current Strategy Forum. Newport, RI: U.S. Naval War College, June 15, 1999.

Kepplinger, Gary. "Decision in the Matter of Lockheed Martin MS2 Tactical Systems." U.S. General Accounting Office Report B-400135, August 8, 2008. https://www.gao .gov/decisions/bidpro/400135.htm.

Kessner, B. C. "Predator B Variant to Challenge Global Hawk for Navy BAMS UAV." *C4I News*, May 1, 2003.

Kilcline, Thomas. "Fitness Report and Counseling Record on Lt Cdr Mark Turner." BUPERS 1610-1. June 7, 2002.

Kinzig, Bill. *Global Hawk System Engineering Case Study*. Wright-Patterson Air Force Base, OH: Air Force Center for Systems Engineering, 2010.

Krepinevich, Andrew F., Jr. *The Military-Technical Revolution: A Preliminary Assessment*. Washington, DC: Center for Strategic and Budgetary Assessment, October 2, 2002. https://csbaonline.org/research/publications/the-military-technical-revolution -a-preliminary-assessment.

Lapins, Maris. Interview by Ernest Snowden. November 19, 2017.

———. "Notes on PBD Timeline." Author's personal papers. January 2004.

———. "Revised BAMS RRP—Issues and Summary of Changes." Author's personal papers. December 18, 2003.

———. "Shaping the BAMS Requirement." Author's personal papers. May 5, 2004.

Leonard, Robert S., and Jeffrey A. Drezner. *Global Hawk and Dark Star*. Santa Monica, CA: RAND, 2002.

"LTV Puts Aerospace-Defense Unit Up for Sale." *Defense Daily*, May 21, 1991.

Miller, Todd. "U.S. Navy MQ-4C Triton Makes Persistent Progress Towards Deployment." *Second Line of Defense*, March 1, 2016. https://sldinfo.com/2016/03/us-navy-mq -4c-triton-makes-persistent-progress-towards-deployment.

Mintz, John. "Aspin Sees Military Promoting Clinton Industrial Policy." *Washington Post*, September 2, 1993.

Muradian, Vago. "Northrop Grumman to Buy UAV-Maker Ryan for $140 Million." *Defense Daily*, May 28, 1999.

"MQ-4C Triton Broad Area Maritime Surveillance (BAMS)." *NavalDrones.com*, September 23, 2016. http://www.navaldrones.com/BAMS.html.

Naval Studies Board of the National Academy of Sciences. *Autonomous Vehicles in Support of Naval Operations*. Washington, DC: National Academies Press, 2005.

"Navy Extends Deadline to May 12 for BAMS Proposals." *Defense Daily*, May 4, 2000.

Nemetz, Matt. "VUP-19 Conducts First MQ-4C Triton Flights." *Planeside*, 3rd Quarter 2018.

"News Breaks." *Aviation Week and Space Technology*, August 21, 2006. https://archive.aviationweek.com/issue/20060821.

"News Breaks." *Aviation Week and Space Technology*, December 18, 2006. https://archive.aviationweek.com/issue/20061218.

Norris, Guy, and Amy Butler. "Head Start: Gulfstream 11 Risk-Reduction Flight Tests Key to Northrop Grumman's Strategy for BAMS." *Aviation Week & Space Technology*, September 24, 2007.

Northrop Grumman. "MQ-4C Triton Completes Critical Milestone to Expand Navy's Maritime Domain Awareness." September 26, 2016. https://news.northropgrumman.com/news/releases/mq-4c-triton-uas-completes-critical-milestone-to-expand-navys-maritime-domain-awareness.

———. "Northrop Grumman Delivers First Operational MQ-4C." November 10, 2017. https://news.northropgrumman.com/news/releases/northrop-grumman-delivers-first-operational-mq-4c-triton-to-us-navy.

O'Guin, Michael. "Shaping the Battle Before It Begins," *APMP Journal*. Fall 2013.

O'Guin, Michael, and Kim Kelly. *Winning the Big Ones—How Teams Capture Large Contracts*. Self-published, 2012.

"P-8 Introduced at Roll Out Ceremony." *Planeside*, no. 1, 2012. https://maritimepatrol association.org/wp-content/uploads/2019/02/MPA_Newsletter_2012_1_compress.pdf.

PEO (U&W) Public Affairs. "MQ-4C Triton UAS Demonstrates New Capabilities During Flight Test." *NAVAIR News*, June 22, 2016.

Phan, N. D. "P-3 Service Life Assessment Program (SLAP)—A Holistic Approach to Inventory Sustainment for Legacy Aircraft." *Proceedings from Tri-Service Corrosion Conference*. Patuxent River, MD: Naval Air Systems Command, 2003. http://namis.alionscience.com/conf/tscc/search/pdf/.

Polmar, Norman, and Kenneth J. Moore. *Cold War Submarines: The Design and Construction of U.S. and Soviet Submarines*. Washington, DC: Potomac Books, 2004.

Powell, Colin. Bottom-Up Review press briefing, September 1, 1993. https://www.c-span.org/video/?49768-1/defense-department-review.

Pralahad, C. K. "Strategies for Growth," in *Rethinking the Future*, Rowan Gibson, ed. Boston: Nicholas Brealey Publishing, 1998.

Proctor, Paul. "New Navy Aircraft Studied." *Aviation Week & Space Technology*, June 28, 1999.

Reade, David. "New Developments." *Maritime Patrol Aviation* 2, no. 4, March 1993.

Richardson, John, Admiral, USN. "A Design for Maintaining Maritime Superiority, Version 2.0." U.S. Naval Institute News, December 17, 2018. https://news.usni.org/2018/12/17 /design-maintaining-maritime-superiority-2-0.

Rivers, Brendan. "UAV Test Beds Ordered by U.S. Navy." *Journal of Electronic Defense*, March 1, 2003.

Roberts, Steven V. "Congress: The Provocative Saga of the $400 Hammer." *New York Times*, June 13, 1984.

Roche, Patrick G. "A FORCEnet Framework for Analysis of Existing Naval C4I Architectures." Master's thesis. Naval Postgraduate School, June 2003.

Rogoway, Tyler. "The Navy Has the Ultimate MH370 Search Tool, It's Just Not Operational." *Foxtrotalpha.com*, March 18, 2014. https://foxtrotalpha.jalopnik.com/why -mq-4c-triton-the-ultimate-mh370-search-tool-isnt-1545912657.

Samit, Jay. *Disrupt You! Master Personal Transformation, Seize Opportunity, and Thrive in the Era of Endless Innovation*. New York: Macmillan, 2015.

Schuhart, Russell. "Prepare for the 'To Be or To Do' Moment." U.S. Naval Institute *Proceedings* 143, no. 12, December 2017.

Shelsby, Ted. "Grumman Says 'Yes' to Northrop." *The Baltimore Sun*, April 5, 1994.

Sherman, Kenneth. "Broad Area Maritime Surveillance Concept Advances." *Journal of Electronic Defense*, January 1, 2001.

———. "P-3 Replacements Proposed." *Journal of Electronic Defense*, October 1, 2000.

Sommer, Geoffrey. *The Global Hawk Unmanned Aerial Vehicle Acquisition Process*. Santa Monica, CA: RAND, 1997.

Spangenberg, George. Oral history transcript, August 31, 1997, 300. https://www.docdroid .com/x9czNLE/george-spangenberg-oral-history-pdf#page=3.

Stratford, Jim. "U.S. Navy Sensor Flight Testing Begins and First MQ-4C Receives Its Wings," Northrop Grumman news release, February 27, 2012. https://news .northropgrumman.com/news/releases/photo-release-northrop-grumman-bams -unmanned-aircraft-system-program-achieves-two-major-milestones.

Strauss, Marc. "Navy Outlines Future UAV Strategy." *Defense Daily*, February 20, 2002.

Swartz, Peter. "Organizing OPNAV (1970–2009)." Center for Naval Analyses Study. Alexandria, Virginia: Center for Naval Analyses, January 2010.

Thomas, Rick. *Global Hawk: The Story and Shadow of America's Controversial Drone*. Self-published, December 2015.

Trimble, Stephen. "Lockheed Protests USN BAMS Award." *Flight Global*, May 5, 2008. https://www.flightglobal.com/lockheed-protests-usn-bams-award/80116 .article.

———. "RQ-4N Spreads Global Hawk Brand to Maritime Patrol." *Flight Global*, February 25, 2009. https://www.flightglobal.com/news/articles/rq-4n-spreads-global-hawk -brand-to-maritime-patrol-323093.

Trimble, William F. *Admiral William A. Moffitt, Architect of Naval Aviation*. Annapolis, MD: Naval Institute Press, 1994.

U.S. House of Representatives, House Appropriations Subcommittee on Defense. Report 107-298, "Department of Defense Appropriations Bill for FY2002." Washington, DC: U.S. Government Printing Office, 2001.

———. Report 107-532, "FY2003 Defense Appropriations Bill." Washington, DC: U.S. Government Printing Office, 2002.

U.S. Senate, Senate Armed Services Committee. Report 107-151, "FY2003 Budget." Washington, DC: U.S. Government Printing Office, 2002.

Van Atta, Richard. *Transformation and Transition: DARPA's Role in Fostering an Emerging Revolution in Military Affairs*. Alexandria, VA: Institute for Defense Analyses, April 2003.

Vesser, Dale. "FY94–99 Defense Planning Guidance, Memorandum to the Service Secretaries." February 18, 1992. www.nsarchive.gwu.edu/nukevault/ebb245/doc01 _full.pdf.

"Vought Aircraft Industries, Inc." *International Directory of Company Histories*, vol. 49. St. James, MO: St. James Press, 2003.

Wakefield, Justin. "A Second Helping of the Last Supper." *Journal of Contract Management*, Summer 2014.

Walker, John. "Why Innovation Must Go Beyond Disruption." *WIRED*, February 2015. https://www.wired.com/insights/2015/02/why-innovation-must-go-beyond -disruption/.

Wood, Robert F., Jr. "Reflections of a Campaign Leader." Author's personal papers, 2008.

INDEX

Figures and notes are indicated by "f" and "n" following the page numbers.

ABOUT THE AUTHORS

ERNEST SNOWDEN is a graduate of the U.S. Naval Academy and a former naval aviator. After active duty, he continued in the Navy Reserve as an aeronautical engineering duty officer until his retirement. As a Navy civilian, he was Business Manager for the Maintenance Policy in the Naval Air Systems Command, then staff assistant to the Deputy Director for Research and Engineering in the Office of the Secretary of Defense. He worked for a succession of aerospace and defense firms in marketing and strategy positions. He is also the author of *Winged Brothers: Naval Aviation as Lived by Ernest and Macon Snowden.*

ROBERT F. WOOD JR. is a retired naval aviator and a veteran of the Gulf War with seven deployments, 4,000 flight hours, 750 carrier landings, and command of an F/A-18 Hornet squadron to his credit. His final tour of duty was for Director, Air Warfare on the staff of the Office of the CNO (OPNAV N88) as the F/A-18 Requirements Officer. Since his retirement, he has pursued a twenty-three-year career in business as a senior executive and consultant in the aerospace and defense industries.

THE NAVAL INSTITUTE PRESS is the book-publishing arm of the U.S. Naval Institute, a private, nonprofit, membership society for sea service professionals and others who share an interest in naval and maritime affairs. Established in 1873 at the U.S. Naval Academy in Annapolis, Maryland, where its offices remain today, the Naval Institute has members worldwide.

Members of the Naval Institute support the education programs of the society and receive the influential monthly magazine *Proceedings* or the colorful bimonthly magazine *Naval History* and discounts on fine nautical prints and on ship and aircraft photos. They also have access to the transcripts of the Institute's Oral History Program and get discounted admission to any of the Institute-sponsored seminars offered around the country.

The Naval Institute's book-publishing program, begun in 1898 with basic guides to naval practices, has broadened its scope to include books of more general interest. Now the Naval Institute Press publishes about seventy titles each year, ranging from how-to books on boating and navigation to battle histories, biographies, ship and aircraft guides, and novels. Institute members receive significant discounts on the Press' more than eight hundred books in print.

Full-time students are eligible for special half-price membership rates. Life memberships are also available.

For a free catalog describing Naval Institute Press books currently available, and for further information about joining the U.S. Naval Institute, please write to:

Member Services
U.S. NAVAL INSTITUTE
291 Wood Road
Annapolis, MD 21402-5034

Telephone: (800) 233-8764
Fax: (410) 571-1703
Web address: www.usni.org